The Institute of B
Studies in Biolog

Ecology and Archaeology

Geoffrey W. Dimbleby

B.Sc., M.A., D.Phil., F.S.A.

Professor of Human Environment,
Institute of Archaeology, University
of London

Edward Arnold

First published 1977
by Edward Arnold (Publishers) Limited
25 Hill Street, London W1X 8LL

Boards edition ISBN: 0 7131 2631 0
Paper edition ISBN: 0 7131 2632 9

Printed in Great Britain by
The Camelot Press Ltd, Southampton

General Preface to the Series

It is no longer possible for one textbook to cover the whole field of Biology and to remain sufficiently up to date. At the same time teachers and students at school, college or university need to keep abreast of recent trends and know where the most significant developments are taking place.

To meet the need for this progressive approach the Institute of Biology has for some years sponsored this series of booklets dealing with subjects specially selected by a panel of editors. The enthusiastic acceptance of the series by teachers and students at school, college and university shows the usefulness of the books in providing a clear and up-to-date coverage of topics, particularly in areas of research and changing views.

Among features of the series are the attention given to methods, the inclusion of a selected list of books for further reading and, wherever possible, suggestions for practical work.

Readers' comments will be welcomed by the authors or the Education Officer of the Institute.

1977

The Institute of Biology
41 Queens Gate,
London, SW7 5HU

Preface

The fact that a book dealing with archaeology is appropriate in a series of biology books illustrates the changes between disciplines which have taken place in the last few decades. Here ecology forms a meeting point for the study of the past environments and those aspects of man's culture and behaviour which are related to his environment. No longer is it possible to study the one without the other; what we have to analyse is a system in which man himself is, and has been for a long time, an ecological factor as well as a member of the ecological community.

Archaeological sites are samples of the past containing direct evidence of past ecosystems in the form of plant and animal remains, soils and, of course, man himself. Progress involves identifying the clues, and drawing conclusions from them, wherever possible testing the conclusions against other independent types of evidence. It is not yet a precise science, and probably never will be, but it is a growing one and a very exciting one to be involved in.

London, 1976

G. W. D.

Contents

1 Modern Archaeology embraces Ecology

1.1 Man's place in the ecosystem

Most introductory books on ecology set about presenting their subject by making statements on general ecological principles: the interactions between organisms and the physical habitat; the concept of community, sere and climax; producers and decomposers; food chains and energy transfer; and hence the concept of an ecosystem. Rarely is man mentioned in this outline, and, if he is it is not as a component organism in the ecosystem, but as an outside influence; sometimes he appears only under the heading 'Applied Ecology'. It is, of course, quite reasonable to establish basic ecological principles in this way, but it can give the impression that there is a 'natural' ecology and an 'artificial' ecology, a view which has some foundation in the history of the growth of the science of ecology, but little justification in practice.

There are very few ecological systems in the world today where the influence of man can be disregarded; if there is no apparent influence now (and this in itself is rare), there often has been in the past. Even man's activities in what we regard as the remote past may have left a mark still recognizable today. Our present landscape in Britain, for instance, shows features which can be traced back even to pre-agricultural man, over 5000 years ago. In areas not affected by glaciation, such as much of Africa, it can be shown that man has been modifying his environment for tens or even hundreds of millennia.

Archaeology in many parts of the world is contributing to a growing understanding of man's inextricable relationship to show that his evolution has been closely bound up with the environment and for much of this time he has had the power to modify his habitat, though perhaps not always intentionally. In those parts of the world where his early evolution took place—usually in the tropics or sub-tropics—the ecosystem itself has developed with him as an integral component of it. Elsewhere in the course of migration he has moved in, and over a period of time has established a niche in the ecosystem into which he came. It is not until the later periods of man's development, and particularly when he had become an agriculturalist, that his influence on the ecosystem became so marked that it could fairly be described as disruptive of the pre-existing ecological condition (in which he may already have been operating, though at a lower level of intensity). In these circumstances for practical purposes it may be useful to distinguish the effects of man from the other factors of the environment, but the fundamental unity of man with the ecosystem remains.

1.2 Man's requirements from the environment

Man is an omnivore; that is, he eats both plant and animal food. However, his dietary habits are very variable throughout this range; some people such as the Eskimo eat almost exclusively animal food, whilst others subsist on vegetable food such as cereals. In the palaeolithic period hunters seemed to exploit mainly the animal food sources, but there is evidence that in a number of regions a change took place early in the post-glacial period to a more varied diet which included smaller mammals, molluscs, crustaceans and plants. The reason for this change is not readily apparent, but it has been suggested that it was precursor of the earliest agricultural systems. Archaeology has a big part to play in providing the evidence for or against such a far-reaching suggestion.

The extent to which man exploits the animal or the plant resources of his environment will determine the degree of ecological influence that he will exert on his habitat. This may be a relatively simple thing to establish for hunter/gatherer man, but once crops are grown and animals raised for food, the ecological impact is much more marked and much more complicated. Such interrelationships will be discussed in more detail in Chapter 7.

Man's demands on his surroundings, however, are not restricted to food. He used shelter and clothing to enable him to live in habitats to which he is not physiologically well adapted. The materials for these purposes are derived from the environment, and the winning of them is one more aspect of man's total impact. Where man is or was practising agriculture, of course, crops of fibre-producing plants were grown for such purposes, and the domesticated animal could be the source of skins and other useful materials.

It is not until relatively recent times that the environment was exploited for inorganic resources (e.g. for tools, weapons, and so on) to such an extent that their extraction seriously affected the local ecology, but the same cannot be said of one other important resource—fuel. Fires were necessary for warmth, for protection against wild animals, and for cooking. The increasing dependence on plant food, to which reference has already been made, required the parallel development of cooking; plant foods, even more than animal foods, have to be cooked to be edible. In some cases cooking is required to remove poisonous or indigestible elements. The traditional fuel is wood, and the search for it has been a factor leading to treelessness of the landscape, especially in semi-arid areas where re-growth is slow. In treeless areas other fuels may be used, such as dung, animal fat and bones, or turf, but wood is always the preferred fuel.

1.3 Man as an ecological force

One of the fundamental principles of ecology is that ecological systems are dynamic; if one of the controlling factors changes, or if one member of the community in some way plays an over-riding part in the ecosystem, the whole system will be influenced; in extreme situations the complete disintegration of the ecosystem may be brought about.

In the light of what was said in § 1.2 above it is apparent that man has for long had the power to alter important components of the ecosystem; in particular he has been able to change the dominant organisms; for instance, through the use of fire he may change a community from one dominated by trees to one dominated by grasses. This concept of dominance is important, for it means more than numerical dominance. A dominant organism literally dominates or determines the complex of the web of life in that community; it also determines such physical factors as evaporation/transpiration, run-off, diurnal and seasonal temperature regimes at soil level, the incidence of direct or indirect light and so forth. Though man may not have been able to change the climate of his habitat, he could certainly change the microclimate, with far-reaching results.

Perhaps the most powerful force available to early man was fire. Man has been able to make fire for 500 000 years in Asia, and at least 100 000 years in Africa and other continents. Much of his evolution has been in climatic conditions that gave rise to vegetation of a type particularly susceptible to burning: grassland, or savanna. Recent experimental work has shown that in areas with a long dry season it is almost impossible to protect the vegetation against man-made fire. In the later periods of man's existence fire has been used deliberately as a tool for changing the vegetation; e.g. the slash-and-burn system of forest clearance. At all times there has been the possibility of accidental fires—fires are easier to start than to stop—and comparison with the present-day incidence of fires in many different types of climate and terrain suggests that fire is much more likely to be started by human agency than by natural factors such as lightning strikes.

Obviously fire, especially if it is of regular occurrence, will have an effect on the vegetation cover. The example was given above of a change from forest to grassland; in regions where fire is a factor of long-standing, species of trees or shrub resistant to burning may have evolved and these are favoured in comparison with fire-tender species in such areas (Fig. 1–1). Fire, however, is not the only influence which man can exert that has a direct impact on the dominant plants. Man has devised tools of stone, and later metal, which can be used for the felling of at least the smaller trees (Fig. 1–2); and other methods can be used to destroy the large ones if desired. The axe was widely used for felling timber for fuel or for constructional purposes. Sometimes certain species were favoured, and

Fig. 1–1 Fire sweeping through African savanna.

this could lead to a marked reduction in their frequency in the surrounding countryside.

Though fire is probably the oldest anthropogenic factor, its effect on the landscape was surpassed by the changes brought about once agriculture was introduced; fire was then just one of a number of practices all designed to replace the natural ecosystems with new ones of man's own choice. Most agricultural crops, whether they are seed crops such as cereals or root crops such as manioc, are plants which need full light for their development. Our own cereals, wheat, barley and oats, for instance, were domesticated from wild species of open communities in the Near East. It follows, therefore, that they cannot be grown under trees, especially in higher latitudes where the sunlight falls at an angle and is therefore less intense. So in forested country clearings have to be made in the forests or, if agriculture is to be more than mere shifting cultivation, whole areas of forest have to be cleared. The intensive raising of animals, too, requires the removal of the tree cover.

Over the long periods of time in which man has been operating secular changes of climate may have occurred. For instance, over the last 10 000 years the climate of Britain has changed from a cold to a temperate one and now is cooling down again. Where the evidence on which these

Fig. 1–2 Dr Js. Iversen felling a small tree with a polished stone axe in wooden haft. (Courtesy Museum of Rural Life. University of Reading.)

deductions are made is botanical, the changes brought about by man may cloud the picture. So often man's influence produced effects which could also have a climatic cause; for instance, is the increased treelessness of the North African coast due to man's influence or to increased aridity of climate—or both? Questions of this sort may require evidence from other sciences beyond the scope of this book.

2 Sources of Evidence

2.1 The past in the modern landscape

Although the landscapes we see today are very largely the product of present-day climate and recent human influences, there are few parts of the world in which it is not possible to find some direct evidence of the vegetation, fauna and sometimes soils (paleosols) of earlier periods. In a country like Britain, which has a history of glaciation, the deposition of material such as boulder clay, wind- and water-lain deposits, or slope wash may cover and preserve remnants of an earlier land surface. Elsewhere massive deposition of material by subaerial erosion, vulcanism and other factors may achieve a similar result. In basin topography, where water collects to form a lake or where peat is forming, sediments are laid down in waterlogged, and therefore anaerobic, conditions. Such conditions retard the aerobic decomposition by micro-organisms, so organic remains tend to be well preserved. Moreover, such sediments are usually well stratified, so that by serial sampling it is possible to analyse the changes in flora and fauna, often over a period of thousands of years. The processes of erosion and deposition, and the accumulation of deposits in basin sites have left even more abundant traces from the post-glacial period, and it is often possible to trace a sequence right up to the present day.

For over a century now studies have been made of the organic remains in such situations. At the turn of the century zoologists were recognizing assemblages of bones, usually from caves, indicating faunas very different from those now in Britain which we now know were associated with earlier phases of the Pleistocene. Botanists were identifying the leaves and fruits of plants and finding that some of these were characteristic of climates quite different from our own; for instance, some were of an arctic character.

2.2 Pollen analysis

The development that changed the course of paleo-environmental research came in the second decade of this century when the Swedish botanist von Post showed that pollen grains, which are preserved in good condition and in great numbers in waterlogged peat deposits, showed fluctuations of species representation through time that could be interpreted in terms of change in the forest composition. Pollen analysis has now been applied from the arctic to the tropics and has vastly expanded our knowledge of past vegetation. Pollen is deposited each year as pollen 'rain' on all exposed surfaces, and is well preserved in anaerobic deposits and other places where decay is inhibited. Thus it may be found in frozen soils, very acid soils, or in desert conditions where the dryness stops microbiological decomposition. Not only has pollen analysis made possible the identification of many taxa not previously recorded, but because pollen is dispersed widely it presents a picture of the vegetation both of the collecting area itself and of the surrounding countryside as well. A fuller treatment is given in Chapter 4.

In the early days of such studies the interest of natural historians was focused on the climate of the past; some workers were mainly concerned to confirm the biblical accounts of human history and evidence of an ancient flood was eagerly sought, but others were trying to build up a chrono-sequence of climatic change. For two or three decades after it came into use, pollen analysis only dealt with tree pollen on the grounds that it was the forest cover that reflected the climate of the past. Following the pioneer work of Iversen in Denmark, however, more attention is now given to the pollen of herbs and shrubs, and it has come to be realized that the pollen record can tell us much more than how the climatically determined forests fluctuated in composition. It can tell us about man's effect on the forests; how his use of the land removed or reduced the forest and what he was doing in the countryside. This theme will be returned to in Chapter 7.

2.3 Evidence from archaeological sites

We can see, then, that these lines of study of the living environment of past times give us a picture of the ecological setting in which man lived and enable us to study his impact on his environment. There is a massive literature on this subject, so this approach will not be discussed in more detail here. An excellent example of the scope and results of such work is Winifred Pennington's *History of the British Vegetation*, a synthesis of pollen analytical work in which the hand of man is clearly demonstrated. It is one thing, however, to be able to see the pattern of changes brought about by man from prehistoric times onwards, but it is another matter to be able to say what effect particular groups of men were having at specific places. As was said in the opening chapter, modern archaeology must put man in his environmental setting; it must see how this setting affected his way of life and how he, in turn, affected his setting. It is not enough to be able to say, on the basis of pollen diagrams from sites unconnected with human occupation, that, for instance, in the Bronze Age the forest was rapidly being pushed back to allow arable farming. We need to know, and it is possible to find out, what effect this Bronze Age group was having on the chalk hills of Wessex, or that Bronze Age group was having on the high hills of the Pennines. Even in one period of prehistory different forms of land use were being practised on different types of terrain, and in some cases the results are still to be seen today.

Fig. 2–1 Buried soil beneath a neolithic long barrow at South Street, near Avebury, Wiltshire.

It is here that the link can be made with archaeological sites themselves. Man creates, artificially, conditions comparable with those produced by natural processes. When he throws up an earthwork, whether it is a neolithic long barrow or an Iron Age rampart, he buries a portion of the contemporary land surface; that is, he creates a paleosol (Fig. 2–1). When he digs a ditch he creates a hollow in which organic detritus will collect and which itself may be covered over by subsequent aggradation. If he digs a well he is creating a waterlogged depression in which a variety of organic remains may become entrapped and preserved. In each case it is protected from later contamination. Not all situations are as simple and clear cut as suggested here, but the principle should be apparent. Whilst ditches, wells and other structures can be useful sources of environmental information, by far the most valuable is the buried land surface itself. Assuming that it has not been truncated (unfortunately it often has been), the old land surface is a preserved microcosm from the past. In later chapters the different types of preserved evidence will be described in more detail and it will be shown that different soil conditions favour the preservation of these differentially. For example pollen is well preserved in acid soils but poorly in calcareous soils; the reverse is true for snail shells. No single buried soil, therefore, will normally contain all possible types of evidence, but any one may yield several types.

Fig. 2–2 Bronze Age plough marks beneath the bank at Avebury, Wiltshire. (Courtesy J. G. Evans.)

Perhaps the most significant fact about buried soils is that the various remains of organic origin—pollen, seeds, charcoal, bones, snail shells, insect remains—represent to a large extent an assemblage of organisms that were living together on or near this site. They therefore represent an ecological unity (or community) associated with man, in contrast to the remains preserved in peat or lake sediment accumulations which, if they represent a community at all, represent an aquatic or specialized one bearing little relation to human pressures. Furthermore, the very soil

Fig. 2–3 Plan of neolithic plough marks beneath a long barrow at South Street, Wiltshire. (Courtesy Leicester University Press.)

itself may provide valuable information about the contemporary ecology and in particular about man's operations connected with it. Plough furrows sometimes show up beneath tumuli (Figs. 2–2 and 2–3) and the distribution pattern of pollen in buried soils can be very informative about the processes of soil working that were in operation, both naturally and artificially (Chapter 3).

2.4 Urban sites

It is natural to think of rural sites as sources of evidence about human ecology, but it should not be overlooked that urban sites can yield a great deal of organic material and this opens up whole new avenues. From Roman times onwards, rich town deposits were laid down, made up largely, but not entirely, of the accumulated refuse from town life. Such deposits may also contain seeds, mosses, or the remains of vertebrates and insects which reflect the urban 'ecosystem'. Investigations of Roman and medieval levels along the Thames in London not only give interesting information about the contemporary conditions of life in a riverside town, but may also give evidence about the Thames itself: how far the tide reached; whether the water was fresh, brackish or salt; or the existence of creeks and tributaries long since obliterated on the surface.

The investigations on urban sites may also shed light far beyond the immediate context. In thirteenth-century Southampton, for instance, food remains extracted from cesspits and pottery vessels showed large concentrations of the pips of grapes and figs which would have been imported, as well as of other foods which may (or may not) have been native. Such identifications can be concrete evidence of trade and demonstrate the extension of urban man's 'environment' to territory far beyond that which he occupies.

Recent work in other Roman and medieval towns, notably York and Winchester, has produced evidence of the unsavoury conditions in which such people as leather-workers or butchers worked. The prevalence of such pests as the timber beetles has been studied in York and elsewhere and our knowledge of their history and the factors governing their distribution in man-made environments has been considerably extended as a result.

2.5 Comparative studies

In a country which has been occupied by man continuously for many thousands of years, it is inevitable that many places contain archaeological sites of several periods. Such a series of sites can provide valuable comparative material. Not only does analysis of the plant and animal remains provide direct evidence of the ecological conditions in successive periods but the development of the soil profile can be demonstrated in a way that is rarely possible by any other means of investigation. This technique has been applied to show the dramatic change in the vegetation and soil of what are now the North York Moors by the successive occupations from the Neolithic onwards (Fig. 2–4; see also Chapter 7).

On a shorter time scale, stages in the construction of an earthwork, such as the enlarging of barrows to include secondary burials, may similarly provide data on the direction of ecological change and of its rate. From such studies it has been found that at certain key periods in human history man has brought about rapid deterioration of his environment. On acid soils Bronze Age farming produced just such accelerated change, which appears to have been followed by a slower progressive deterioration which still goes on.

Fig. 2–4 Reconstruction of landscape based on pollen analyses from buried soil beneath the late neolithic chambered cairn at Great Ayton, Yorkshire, North Riding.

Upper: panorama photograph of modern landscape
Lower: interpretation of neolithic landscape.

3 Nature of the Evidence: the Physical Habit

3.1 Forces of weathering

The mineral surface of the earth is continually being subjected to the force of climatic and organic weathering. The various agencies at work give rise to products of different physical characters. Frost working on a rock face, for example, produces coarse angular fragments through the pressure exerted by water freezing in the cracks and crevices; this is the source of the screes which we see at the base of rock faces, as in the well-known screes overlooking Wastwater in the Lake District. Such screes may be added to during hard winters even today, but they mostly date back to the late-glacial period over 10 000 years ago.

In tropical areas, especially under desert conditions, the intense heat of the day may give way to near-freezing temperatures at night, and the consequent thermal expansion and contraction of the rocks can cause exfoliation, that is, the flaking off of the superficial layers. This, too, will give rise to coarse angular deposits, but the shape of the fragments will be different from that caused by frost action.

In temperate and tropical climates with adequate humidity chemical solution of rocks and deposits takes place. The more soluble minerals are removed and clayey or loamy deposits may be left behind. The percolating waters become enriched and on evaporation may deposit their dissolved minerals. In temperate regions stalactites and stalagmites are found in caves in limestone rocks, or surface layers of calcium carbonate may be laid down as travertine or tufa. In the tropics where the high temperatures result in the almost complete disappearance of the soluble calcium salts, silica can come into solution and may be deposited in the form of silcrete.

Such considerations as these have frequently been of great archaeological significance, especially on early prehistoric sites. Palaeolithic—Old Stone Age—remains are commonly found in caves, and the detritus making up the floor of the cave can, through its physical characteristics, be interpreted in terms of factors such as frost or humidity. Bones, both human and animal, together with man's artifacts, may be stratified in such deposits and can therefore be allocated to the appropriate climatic condition.

Palaeolithic remains are also found in the deposits making up the terraces of the larger rivers, and such terraces can be attributed to certain phases of the glacial sequence. A glacial period will produce a great deal of frost-weathered detritus which is carried away by the rivers as the land thaws at the beginning of the following warm phase, eventually being dropped by the river over its flood plain. There are archaeological problems associated with such situations, for the artifacts, e.g. hand axes, were not necessarily dropped by man on the surface of the terrace on which they were found; they might have been dropped at an earlier time and then carried by the river as part of its gravel load. There are ways of separating these two possibilities, but for these reference should be made to archaeological books.

3.2 Soils and their development

Deposits produced by the physical weathering of rocks will begin to develop a soil where they are exposed to the influence of atmosphere and organisms, particularly plants. Percolating water moves finer particles downwards through the material, until they accumulate perhaps as much as a metre below the surface. Rainwater is slightly acid through the solution of carbon dioxide from the air, and this acid solution differentially dissolves certain materials such as carbonate of lime (in chalk, limestone or other base-rich deposits) and other relatively soluble substances. Much of such dissolved material may pass out of the soil into the deep drainage, but some may be retained in the lower soil levels. But in a situation where vegetation is growing, the percolating water will also be enriched with organic acids from the breakdown of plant and animal remains, and these acids increase the washing-out effect (leaching) of the soil water. Under certain conditions this can lead to a removal of almost all the soil constituents except silica, leaving a bleached sand beneath the surface humus. The deep layers of the soil are heavily stained with iron and humus, and contain other substances washed down from above (this is a podzol soil—see below). In tropical soils intense evaporation may lead to the *upward* movement of water, so that the *surface* becomes enriched in comparison with the lower layers (ferralitic soils).

As a soil develops, therefore, it tends to become differentiated into horizons, having started as a uniform deposit. It is not possible to discuss the question of soil genesis in more detail here, but two important principles follow even from what little has been said. Firstly, horizon differentiation develops progressively until a profile is produced which does not show any detectable change with increasing time. The profile is then said to be mature and represents an equilibrium between that particular deposit (parent material), the vegetation and the climate (and in some cases the topography, too). In temperate regions soils may take 1000–5000 years to reach maturity.

The second principle is that a change in one or more of the soil-forming factors may result in a different type of soil profile. Thus a climatic change—which could well occur in a period as long as 5000 years—may cause new features to appear in an already well-developed profile; an increase of precipitation, for instance, would lead to intensified leaching. From the archaeological standpoint the important fact is that man almost always changes one of the soil forming factors, namely vegetation, so it is to be expected that his impact may be reflected in the soil profile (see Chapter 7). In so far as the mature soil is in equilibrium with the environment, any soil influenced by man will be at a lower level of organization (unless he has increased the fertility of the

Fig. 3–1 Modern erosion of moorland, North York Moors.

system by the addition of extraneous materials such as manures or fertilizers); in extreme cases of soil exploitation it may not even be possible to establish a new equilibrium, and progressive deterioration may take place. Using the mature soil as a yardstick, such lower levels of organization or fertility can validly be described as degraded. Where processes of degradation are not stabilized they can result in collapse of the ecosystem. This may even be manifested in erosion (Fig. 3–1; see also § 3.3 below), or in such chemical condition as salinization, that is, the accumulation of salt at the surface as a result of inefficient irrigation practices in an arid climate. Both conditions have bedevilled man's use of the land from the earliest days of agriculture, and their effects can be detected in the archaeological record.

Material for pedological study is found in many types of archaeological sites. The mounds of tumuli, and earthworks of many other types, are thrown up on top of a contemporary soil (paleosol) (Fig. 2–1). In some cases the soil may have been seriously disturbed; sometimes it has clearly been truncated, that is, the surface horizons have been removed, perhaps by erosion, or perhaps by man's own operations. But even a truncated soil may be worth studying, especially if the characteristic lower horizons are still intact. Even within the earthwork itself useful studies can often be made. If a mound or rampart has been heightened, the original surface, now buried, may show some evidence of profile development and this might imply that there was a considerable time gap before the secondary addition was made. Many earthworks include as part of their bulk turves cut from the surrounding land surface. These may be thick enough to give valuable information about the adjacent land surface, and are particularly useful where the old land surface beneath the earthwork has a truncated soil profile.

3·3 Soil movement

As indicated above, the movement of soil material, that is, erosion in some form, can result from man's interference with the ecosystem. Erosion is a process which is taking place all the time. The forces of wind and water, perhaps assisted by gravity on slopes, are ever-present. Under a continuous cover of vegetation, however, the impact of these forces is minimized, so that the rates of soil movement are slow. Nevertheless, massive movements such as landslides can and do occur, though rarely, even under a cover of vegetation. If, however, the plant cover is broken, or removed altogether, the control of the rate of erosion is broken and soil movement may take place. It is sometimes said that man accelerates erosion, but it would be more accurate to say that he merely removes the brakes and allows the natural erosive forces to act unhindered.

The effect of these forces depends not only upon their intensity, but also upon the nature of the soil material. A deposit of gravel will not be disturbed by the wind, and only by water if a flood of high velocity develops. On the other hand, a soil which has developed in loess will be particularly susceptible to wind erosion. Loess itself is a wind-carried deposit laid down by the action of high winds working on the unsorted deposits left behind as ice sheets retract. Many soils will slip downhill under the influence of gravity, particularly if they become waterlogged or overlie an impervious clay substream.

Evidence of soil movement frequently turns up on archaeological sites, and indeed in some cases it is the mass movement of soil material which has preserved the sites. A not uncommon example of this is the burial of a site underneath a sand dune; such sites may be of almost any age—I have worked on mesolithic, neolithic and Bronze Age sites in such situations. Amongst other things, good soil profiles are often preserved beneath

Fig. 3–2 Forest soil buried beneath sand dune, Isles of Scilly. The upper dark line is a temporary soil surface developed when the dune was stable, later covered by further sand blow.

dunes (Fig. 3–2). The dunes themselves may have arisen from much later disturbance of ecosystems on a vulnerable substratum.

In hilly country hill wash often takes place on a massive scale; the clearance of primary forest by neolithic or Bronze Age farmers has led to sheet erosion when the natural vegetation has been removed over a large area. In situations where forest regeneration is rapid, so that clearings are quickly covered over, little erosion seems to have occurred. It is commonly assumed that erosion is associated with a wet climate, but in fact some of the most serious erosion takes place in semi-arid climates where the plant cover is thin and sudden violent storms, especially in the dry season, can cause flash floods of enormous erosive power. Even in this country examples could be quoted of the movement of massive quantities of soil and rock as the result of a single storm.

Nevertheless, the gradual movement of soil downhill, perhaps associated with tillage, can produce a major redistribution of topsoil over a period of time. The build-up of lynchets, the banks which form between Celtic fields, is a good example.

Archaeological sites may receive soil material as well as lose it. The burial by dunes has already been mentioned, but in areas of sandy soil sites of various ages can be found beneath layers of blown sand perhaps a few centimetres thick. This is how mesolithic sites on heathland are often preserved. In later monuments ditches are sometimes filled up with windblown material. This does not imply that the climate was any windier or drier than now, but merely that sufficient vulnerable soil was exposed at the time when drying winds were blowing—as is common in the spring.

3·4 Solution of subsoil

Atkinson has drawn attention to the fact that on chalk soils it can be shown that the old land surface buried beneath a prehistoric monument may stand as much as $\frac{1}{2}$ m higher than the present land surface adjacent to the monument. He suggests that this is due to the solution of the soluble chalk which has taken place in the millennia since the monument was built. That this is not the whole explanation is shown by the fact that a similar, though smaller difference is found in siliceous gravels, which are insoluble. It should also be remembered that if the chalk surface is dissolving away as rapidly as Atkinson believes, the surfaces of the earthworks themselves may be being reduced at a similar rate. It may be that cultivation has contributed to the loss of soil, but it is salutary to be reminded that such changes have gone on; some are now operating at much accelerated rates. Not only does cultivation affect the adjacent land but today it is threatening many monuments themselves. Modern ploughing frequently goes right over round barrows, and in places like the Yorkshire Wolds there are examples of barrows which 50 years ago stood nearly 2 m high now being scarcely detectable.

4 Nature of the Evidence: Plant Remains

4.1 The composition of plant material

We now turn to actual remains which have persisted since the date of the archaeological site. As every gardener knows, plant material quickly decays; woody material persists longer than the softer tissues but it is only a matter of time before it goes too. Why is it, then, that plant material can survive on archaeological sites for very long periods of time?

First of all it must be said that most of it does not. What we find is that which has been preserved because of some feature of its composition combined with the micro-environmental conditions in which it has lain. Generally speaking the living parts of the plant cell contents are not preserved; they are either soluble or readily hydrolysable and are quickly converted by micro-organisms. In the absence of moisture even material such as starch can be preserved, as in tubers found in rock shelters in South American desert areas, but this is exceptional. Usually it is the materials which give mechanical strength which are preserved: the cellulose of the cell walls, or the cellulose and lignin of the xylem and extra-stelar mechanical tissue such as pericycle fibres.

4.2 Conditions leading to preservation

Decomposition is mainly by aerobic bacteria and fungi, dependent upon a supply of oxygen, adequate moisture and a suitable temperature. Furthermore the bacteria (including the related Actinomycetes) are sensitive to acidity. Where all these factors are at tolerable levels, decomposition of all plant products will be rapid, but where one or more become limiting decay will be hindered; that is, there will be a degree of preservation.

There are various ways in which conditions on archaeological sites may promote preservation. Extreme aridity has already been mentioned; the splendid preservation of organic materials in the tombs of ancient Egypt is another example. The organic materials may be severely distorted by shrinkage but they are not decomposed. Extreme cold can also produce magnificent preservation, as for instance in the ice-bound Pazyryk tombs in the Altai mountains in Central Asia. Here both animal and plant material was preserved, the textiles even retaining their coloured designs.

The most frequent situation in which organic material is preserved, however, is when waterlogged. The effect of waterlogging is to exclude free air and therefore oxygen; the only form of decay which can proceed is anaerobic, but anaerobic bacteria are not the main agents of decay of cellulose and lignin. Consequently preservation is good. Occasionally it appears that anaerobic conditions can occur in non-waterlogged situations. Under Silbury Hill, the largest man-made mound in Europe, preservation of plant material, even unlignified cellulose, was very good. A likely explanation is that the large amount of turf incorporated in the mound quickly used up the oxygen through aerobic decay, and that as further diffusion into an earthwork of this size was negligible, no further decay took place.

The biochemical constitution of some plant materials may have bacteriostatic effects. In particular tannins and the complex polyphenols have been shown to act in this way. For instance, deposits containing large quantities of tannin-rich material such as oak bark, sometimes found associated with tanneries, may show good preservation of material normally susceptible to decay.

4.3 Macroscopic plant remains

Preserved plant remains can usefully be considered under two headings—macroscopic and microscopic. The most commonly found macroscopic material is wood. In anaerobic deposits it may be well preserved, though in some circumstances tissue disorganization does occur. Very commonly, however, wood occurs as charcoal; that is, it has been reduced to elemental carbon by incomplete combustion. Elemental carbon is indestructible by micro-organisms or other soil processes so charcoal is preserved indefinitely. The remarkable thing is that the combustion process does not destroy the microscopic structure of the wood. Cell walls, lignified thickenings and even bordered pits and multiple perforation plates in the vessels are beautifully preserved. It is therefore possible to identify charcoal almost as easily as uncharred wood.

Other macroscopic remains are leaves, stems and bud-scales, though it is not always possible to give a firm identification of such remains. The mosses are sometimes found well preserved and it is usually possible to identify them to species level. Seeds are an important item on many archaeological sites and can be extracted in large quantities by flotation techniques. On sites from the agricultural periods they are often carbonized; in some cases this makes identification more difficult, but by no means always. On anaerobic sites seeds may be well preserved even though not carbonized. Under Silbury Hill, mentioned above, seed preservation was remarkably good, even the natural colour being retained (Fig. 4–1). Seeds are particularly useful because in many cases they can be identified to species level, whereas wood and pollen can rarely be taken beyond genus. Moreover, many seeds are from plants associated

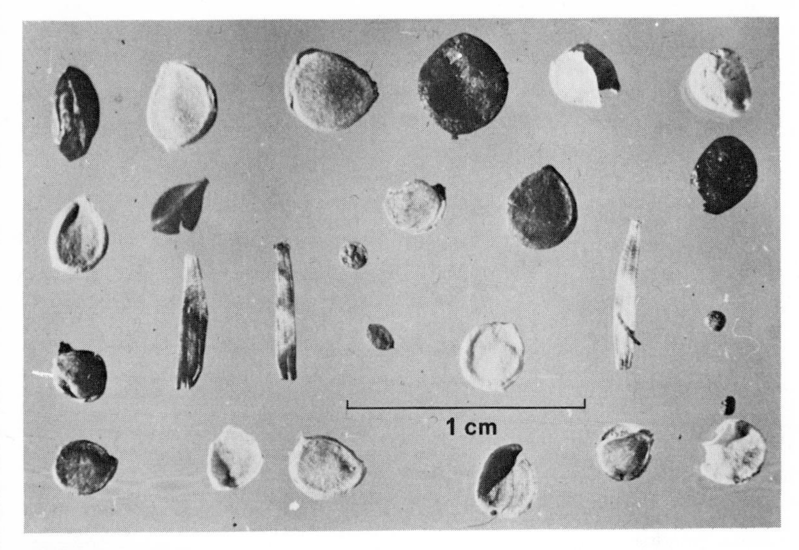

Fig. 4–1 Seeds from the old land surface beneath Silbury Hill, Wilts.

with agriculture and can therefore be useful indicators of the agricultural operations being practised.

Alongside seeds one should mention fruits, both in the botanical and horticultural senses. Frequently such remains are of more economic than ecological significance, but they can give important evidence about man's domestication and early use of fruit.

4·4 Microscopic plant remains

Of the microscopic plant remains pollen is by far the most important. The walls of pollen grains are impregnated with sporopollenin, which in some of its forms is about the most bacterially resistant plant material known. Why pollen grains should be so endowed, when their function is so ephemeral, is not obvious. Furthermore, there is a great diversity of structure of pollen grains, with strong systematic characters, so making identification to family, genus and sometimes species possible (Fig. 4–2). Again it is a mystery why there should be such morphological diversity in a part of the plant which serves basically the same function in all species, but upon these two baffling features the whole science of pollen analysis depends.

Pollen is released into the atmosphere in large quantities in the flowering season especially from wind-pollinated trees and grasses, and may be carried long distances on the wind. Nevertheless, the bulk of the

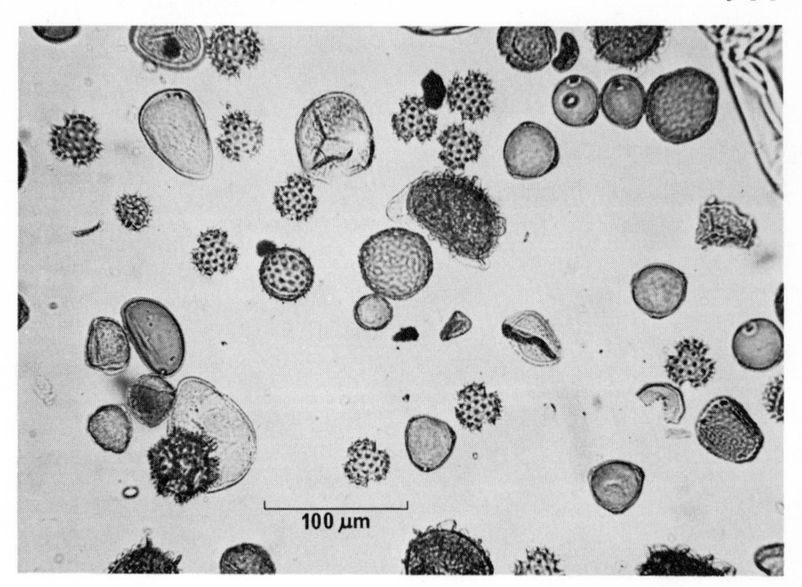

Fig. 4–2 A selection of modern pollen grains and fern spores.

pollen settling out at any given point (pollen rain) comes from local vegetation; consequently, the statistical representation of pollen transported from far off is generally insignificant, though it is important that the appropriate form of numerical representation should be employed. Pollen rain falls on all surfaces—soil, roads, houses, open water, and so on—but it only accumulates in a usable way in certain conditions. Ideally, the pollen analyst requires to know what was the composition of the pollen rain at some given time in the past, and for this two conditions must be met. In the first place, he must be sure that all the pollen he is extracting from his sampling point must be of the same age: it must not be contaminated with earlier or later pollen. Secondly, the pollen he extracts should as far as possible reflect the proportions of the different species present in the pollen rain; that is, any alteration of the proportions by differential decomposition—for not all pollen grains decay at the same rate—should be minimal.

The most favourable conditions are met with in waterlogged deposits which are building up by steady accumulation, such as lake muds or peat bogs. Here not only is aerobic decomposition reduced to a minimum, but all the pollen at any one level is coeval. From such deposits we have learnt a great deal about the changes in landscapes over the past millennia, and we can often recognize the hand of man at work in these changes. A limitation of this approach, however, is that such sites were, by

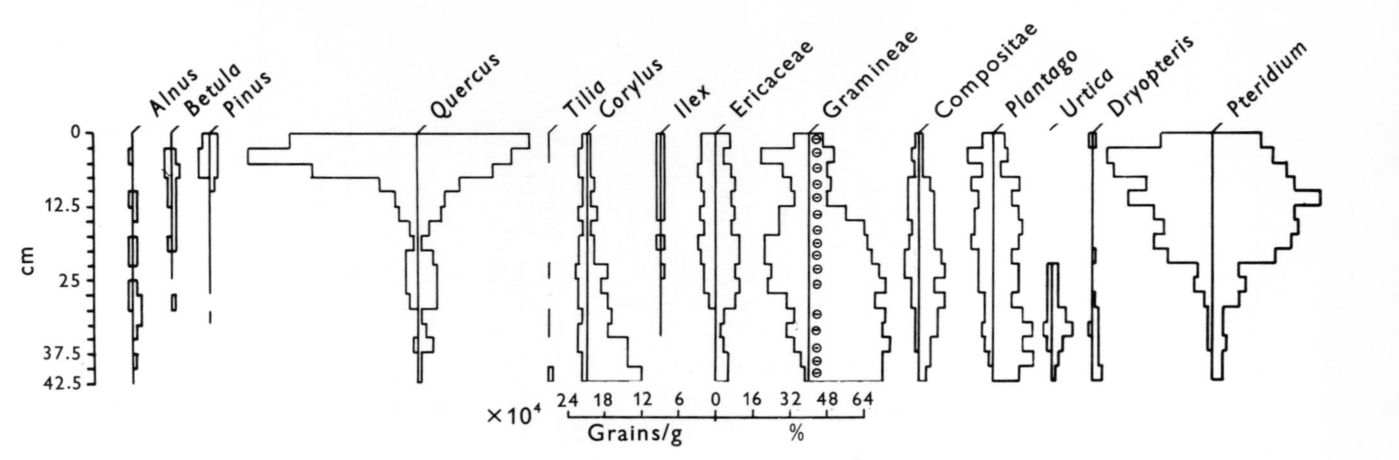

Fig. 4–3 Pollen diagram from modern soil in Matley Wood, New Forest, showing sequences caused by downwash of pollen. (For each taxon the left-hand histogram represents absolute frequency and the right-hand one percentages of total pollen and fern spores.)

their very nature, not used by man; he was normally living in drier places in the neighbourhood, so the evidence is indirect. It was shown in Chapter 2 that archaeological sites can provide direct evidence, and fortunately pollen analysis can be applied in these situations. It is most applicable to sites on acid soils where the acidity has inhibited microbiological process of decay, but it can also be applied on some other types of site. If the old land surface under an earthwork can be recognized it may yield a sample of pollen which is more or less contemporary with the earthwork. As pollen lies in the soil it gets carried down the profile by downwash and by the action of soil animals where these are present. It follows, therefore, that the pollen in the soil surface has not been long in the soil and can be related to the contemporary vegetation.

Fig. 4–4 Natural cast of multiple perforation plate, probably of alder, from a soil under early urban deposits, Southwark.

The downwash of pollen is itself sometimes useful, for pollen which has lain longest in the soil will tend to lie deepest; consequently it is sometimes possible to read the sequence of vegetation change, culminating in the pollen spectrum in the surface layer (Fig. 4–3). Where vigorous mixing of the soil takes place, as for instance by earthworms, such patterns are destroyed. However, earthworms occur in soils which are circumneutral or alkaline (see Chapter 5), conditions in which pollen preservation is poor or non-existent.

Apart from pollen a great variety of microscopic plant debris may appear in preparations made for pollen analysis: fungal and moss spores, fragments of tissues, hairs, and so on. Until more work has been done on these, few of them can be used diagnostically. One exception is the multiple perforation plates (see § 4.3 above)—or rather organic casts of them—which turn up in some soils, and which may correlate with other evidence. For instance, in soils where alder or hazel was growing the pollen of these species is frequent, and so are these characteristic multiple perforation plate casts. The pollen tells us that a certain tree species was growing nearby, but the perforation plates, derived from the decay of root or stem wood, indicates that it was probably rooted on that site (Fig. 4–4).

In making pollen preparations part of the rather violent chemical process (which leaves the pollen unscathed) is the use of hydrofluoric acid to destroy the soil silica. Certain plants contain a good deal of silica, and in the grasses and sedges this can take the form of phytoliths—small bodies of opaline silica laid down in the cells of the epidermis (Fig. 4–5). These may have some diagnostic value and can be of archaeological significance: for instance, layers of ash containing large quantities of phytoliths were presumably derived from fires of grass, grass turves or dung and not of wood. However, such evidence is destroyed in the preparations made for pollen analysis, so special techniques avoiding the use of hydrofluoric acid must be used if the phytoliths are to be extracted and studied.

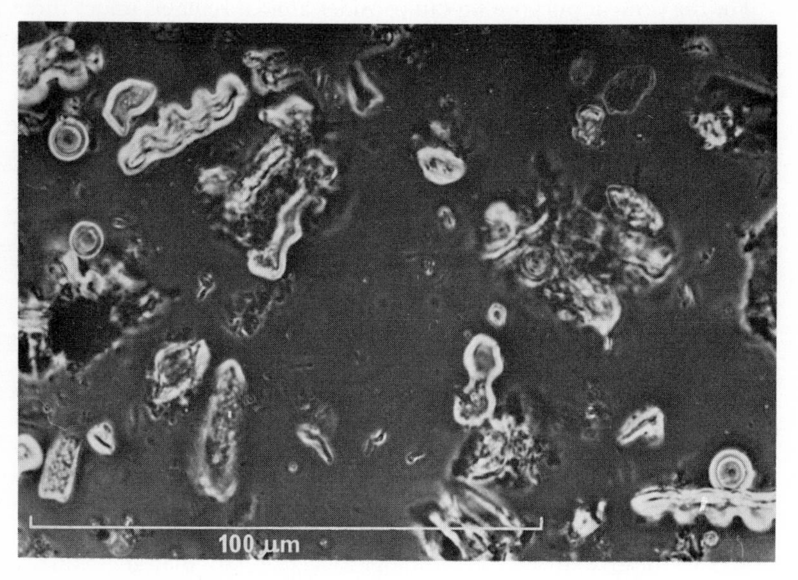

Fig. 4–5 Phytoliths from Dun Ardtreck, Skye (Iron Age).

5 Nature of the Evidence: Animal Remains

5.1 Composition of animal material

The soft parts of animals are biochemically much the same as the soft parts of plant tissue and like it are seldom preserved. Animal tissues have delicate cell membranes and consequently the soma does not persist in a recognizable form as do plant bodies. On the other hand, animals develop their own form of mechanical support, the skeleton, which does not have its parallel in the plant kingdom, and it is this skeletal structure which provides virtually all the direct evidence of the past occurrence of animals. Taking the animal kingdom as a whole there is considerable diversity in the types of skeletal material—more so than in plants. In some cases it is mainly mineral, as in the calcareous shells of molluscs, and as such is not a suitable substrate for microbiological decay. Consequently such material may persist in conditions where microbiological activity is great, giving us evidence about early faunas and, by inference, early ecosystems, in situations where organic evidence is lacking.

For the present purpose we can consider animal remains under three headings: vertebrate skeletal remains; invertebrate remains, with exoskeletons of chitin (Arthropoda) or of a mainly mineral nature (Mollusca); and thirdly the soil animals. The last may leave only indirect traces, e.g. stone-free soil sorted by earthworms, though in some cases, e.g. mites, exoskeletal remains also persist. There are a few other zoological residues which do not fall into these categories (e.g. eggs of internal parasites) and these will be mentioned in the appropriate contexts.

5.2 Vertebrate remains

Only in exceptional circumstances (e.g. the bog corpses of Denmark and elsewhere) do the soft parts of vertebrates survive. More usually it is the bone skeleton and associated structures such as horns, and antlers. Bone has a dual composition, a hard mineral structure of apatite (a calcium phosphate) through which runs a proteinaceous material called collagen. Being of widely different nature, these two materials are preserved under different conditions. In normal soils the collagen is gradually destroyed by microbiological action, so that the bone is progressively depleted of nitrogen, whilst remaining hard and 'bony'. (This was the basis of one of the methods by which the Piltdown fraud was detected.) At the other extreme, as in an acid bog, the apatite may be dissolved, and bone is seldom preserved; this was the condition of the bog corpses referred to above. In many situations both processes are occurring at the same time, but as far as the archaeologist is concerned the most useful rule of thumb is that in calcareous sites bone is well preserved, that is, it retains its shape and is hard and handleable, whereas in more acidic soils it becomes very friable and often has disappeared altogether. Calcined bone, that is bone which has been burnt, seems to be more resistant to chemical solution; it is, of course, purely mineral, the organic fraction having been destroyed by calcining.

The archaeologist is concerned with bones in many ways. If the bones are human they are vital evidence about the people he is studying: their stature, health, customs and so forth. If they are the bones of domestic animals he may learn a great deal about domestication and breeding, methods of stock farming and butchery practices. In terms of ecology all this may be very relevant, particularly when a pastoral economy can be inferred. But the larger herbivores are not the best indicators of the ecological condition of the range; for instance, it is hardly possible to tell whether the grazing pressure was low or intense in any given area. Some carnivorous animals, such as the fox, range widely and therefore are not good ecological indicators, but many of the smaller vertebrates have more restricted habitats: examples are moles, hedgehogs, the various voles, amphibia (frogs, toads and newts), lizards and bats. Whilst animals move about, and are therefore less tied to a given area than are plants, they often stay within the bounds of certain ecological systems; appropriate small vertebrate faunas can indicate the presence of grassland, heath, woodland, marsh, etc. On some archaeological sites it appears that the bones have been brought to the site, perhaps by birds of prey; if so, then the location of the habitats represented may be more remote than would be the case if the animals had reached the site of their own accord.

5·3 Invertebrates

The two most important groups of invertebrates for our purposes are the molluscs and insects. The phylum Mollusca includes a great variety of animals, especially marine ones such as bivalve and gastropod shellfish, cephalopods and also freshwater and land snails and slugs. The shellfish can be important in archaeology as a food source, and their shells were used for decoration or even currency, but for the purpose of our present subject the land snails are the key group (Fig. 5–1). Nor is it the familiar garden snail (*Helix aspersa*) which is important, but the inconspicuous little snails, mostly only a few millimetres long, which inhabit the soil in base-rich habitats (they require the lime for the construction of their shells). As may be imagined, such animals are not very mobile, and some have quite strong ecological preferences; that is, they are good ecological

Fig. 5–1 Land snails from the soil beneath the Devil's Dyke at Newmarket (Roman).

indicators. Some like moist habitats, some short turf or open ground; others prefer woodland or rocky scree; yet others are temperature sensitive, preferring cold or warmer conditions as the case may be. The British molluscan fauna is well known, and though their taxonomy is a job for the expert they have the advantage that for identification the shell does not have to be intact. As long as the apex, and in some cases the mouth, of the shell is present, identification is possible; moreover, population studies can be carried out much more easily than on bones, where one individual may contribute many bones to the assemblage.

Mollusc shells consist mostly of calcium carbonate and therefore are preserved in neutral or alkaline soils. In the fresh condition they have an outer proteinaceous skin, which disappears in the sub-fossil state, though I have seen specimens from acid waterlogged sites where only this proteinaceous skin remained, the calcareous part of the shell having been dissolved away.

The insects are a very numerous and varied group, many of them of high mobility. The Coleoptera (beetles) are better preserved than other groups of insects because of their robust exoskeletons and it is from them that most of our environmental information comes. Some beetles have been found to be closely associated with certain temperature regimes in different stages of the Pleistocene.

It is the chitinous exoskeleton of beetles that is preserved, and from the details of the wing cases, body segments, appendages and genitalia identification can usually be made to species level. As the exoskeleton is organic, one expects to find the best preservation in anaerobic, particularly waterlogged conditions. This is indeed the case, and the richest collections of insect remains have come from such situations as well shafts and ditches. Such collections of insects are referred to as 'death assemblages' signifying that they were not all living together in the same habitat but that their bodies came to be preserved in the same place. Such a collection may contain the odd representatives of habitats which must have been remote; in any case, a single specimen of a species cannot be of great ecological significance. But where a number are present which share a common habitat, then an ecological interpretation becomes justified.

Insect remains are also found in other contexts, especially in the foul conditions found on urban sites in medieval times. It is in cesspits and similar features of such towns that we also get evidence of other invertebrates, particularly the eggs of internal parasitic worms, which reflect the health or otherwise of the townspeople.

5·4 Soil fauna

The animals which dwell in the soil are of particular interest for two reasons: in the first place they can tell us a great deal about the biological activity in an old soil, and this may stand in strong contrast to the biological activity of soils in comparable situations today. Secondly, the activity of animals in the soil may disrupt the distribution of other forms of evidence (see § 4.4 above). Moreover, as Darwin showed nearly a century ago, the activity of earthworms can actually bury archaeological levels.

Evidence of past earthworm activity can be obtained indirectly from features of the soil profile or from the distribution of pollen in the soil profile. As we shall see in Chapters 6 and 7, sites which today have infertile soils may in prehistoric times have had soils of relatively high fertility and certainly with a high biological activity, as reflected by the soil animals. Earthworms are not the only animals which can give us information; Dutch work has suggested that in some heathland soils faecal pellets of micro-arthropods (e.g. mites) characteristic of oakwood soils may still be found, indicating the dramatic change which man has brought about in the ecosystems of such areas.

6 Synthesis

6.1 General

In the previous chapters we have seen the sort of evidence about past environments that can be preserved in archaeological sites. Because the chemical nature of such remains varies so widely it is not to be expected that any one site will include all types of evidence. A calcareous freely-drained site may contain bones, molluscs and charcoal, but little or no pollen or uncarbonized plant tissues. An acid, freely-drained site may contain abundant pollen, charcoal and perhaps have a highly significant soil profile, but will not contain bone or molluscan shells. A waterlogged site may contain macroscopic remains of plants, pollen, insect fragments, but few land snail shells and no useful soil profile. Consequently, the interpretation of those remains which are present is rather like detective work, using what clues are available.

It must be borne in mind that our object in this work is to establish the ecological condition of the landscape in which the site was set. It easy to become sidetracked; the identification of organic remains often provides the archaeologist with valuable information about cultural matters; e.g. the seeds and other remains contained in urban cesspits may be valuable evidence of dietary habits. The investigation of such things usually and quite properly falls to the biologist but they are often classified by the archaeologist as 'environmental', which may not be the case. On rural sites the distinction may be blurred. A collection of cereal seed on a site is of cultural significance, but it may also contain the seeds of weed species which are useful pointers to the ecological condition of the land.

6.2 Use of indicators

In discussing the different types of biological evidence in Chapters 4 and 5 attention was drawn to the fact that some organisms are better ecological indicators than others. Plants with a wide range of ecological tolerance are less useful than those which have a close affinity with certain habitats—what the plant sociologist would call 'fidelity'. Similarly, some animals are more narrowly confined to certain habitats than others, and so are better indicators. A large carnivore will hunt over a varied range of habitats, but a small rodent may be confined to one specific habitat type.

By using the indications of the plant and animal species, broad habitat features can be recognized. For instance, from pollen analyses or molluscan analyses we may deduce that the contemporary environment was wooded because there was a predominance of woodland or shade-tolerant species. Or, as is often the case, both shade-dwellers and light demanders appear in the same assemblage, showing that a patchwork or mosaic of wooded and open ground occurred in the vicinity of the site. Not infrequently it is found that the pollen in the soil of a prehistoric site shows marked contrast to the vegetation of today, as examples below will show, and these differences are usually reflected in a contrast between the prehistoric soil profile and the modern one. In fact this makes the point that wherever possible more than one line of enquiry should be pursued; it adds conviction if the pollen, seeds, and mosses, for instance, all point to the same conclusions. There are circumstances, particularly on calcareous soils, where the pollen and molluscs tell different stories. Assuming that their ecological requirements have not changed over the last few thousand years, this may mean that the two sources of evidence relate to different phases of the site's history. In a calcareous soil pollen is much more ephemeral than mollusc shells, so this may be a key to the problem. Such discrepancies emphasize the need for a synthesis of several lines of evidence rather than relying on only one.

6.3 Sampling and contamination

Before giving some specific examples of the application of these principles a word needs to be said about sampling. There is not space here to go into details of sampling techniques, which must vary according to the size and abundance of the objects to be sampled and their distribution both laterally and vertically in the site. Pieces of charcoal clearly offer a different problem of sampling from, say, pollen grains. Again the sampling of bones poses different problems. There are, however, some principles of importance in all sampling procedures. If we are to be able to interpret the biological evidence so that it gives a picture of the environment at certain time in the past, it is essential that the samples we analyse are contemporary with the archaeological evidence on which the dating is based. This correlation of material is primarily the responsibility of the archaeologist excavating the site, but the environmentalist has a duty to check it for himself. He is perhaps more qualified to observe possible disruptive features of the environment than is his archaeological colleague. Re-working of levels by natural processes such as worm activity, wind and water, animal burrowing, and so on, can be sources of error in relating samples to the correct archaeological level.

Errors can also creep into some forms of sampling through contamination. This applies more to the small and often invisible objects such as pollen grains and seeds than it does to larger objects like bones. Sampling tools must be maintained in a clean state; samples should not be handled by dirty fingers and they should be packaged so that no extraneous contamination can reach them.

6.4 Examples

(a) Silbury Hill, Wiltshire

Silbury Hill has challenged archaeologists for generations and still does so. In 1968, Professor R. J. C. Atkinson tunnelled to the centre of the level of the old land surface, and though little of archaeological value emerged, except the establishment by radiocarbon dating that it was a late neolithic structure, environmental material was well preserved (see § 4.2 above). The *pollen analysis* showed poor representation of tree species and even of hazel; grasses and weeds predominated, which, being light-demanders, confirmed that the country was open. Moreover, the weed pollen included considerable amounts of ribwort plantain (*Plantago lanceolata*), which is characteristically, though not exclusively, found in pastureland. The absence of cereal pollen and weeds of bare soil suggest that arable farming was not being practised in the immediate vicinity.

Among the *macroscopic plant remains* on the old land surface the leaves of grasses could be recognized; they still had some green coloration when first excavated. It was noticed that the leaves were square-ended, not pointed, suggesting that they had been grazed by stock. The old land surface also yielded *seeds* (Fig. 4–1), not in great abundance but in a very good state of preservation; they are typical agricultural weeds, many of them weeds of pasture. A number of mosses were also identified and turned out to be species of grassland.

Insect remains were not abundant, but fragments of ants, again of a species characteristic of grassland, were found. These ants were at the winged stage, which shows that some of the constructional work took place in high summer. Though the old land surface itself was non-calcareous, being based on clay-with-flints, some of the turves of the mound had been cut from the adjacent chalk land, and these contained *mollusc shells*, which on analysis proved to be predominantly grassland species. The *soil profiles* of both the old land surface and the turves were compatible with this interpretation.

A small amount of *charcoal* was found; this was twiggy material and proved to be hazel.

Such evidence in this case matches up and gives a picture which probably reflects the actual land use at the time the mound was constructed. This case indicates the way in which the various clues from such a site can be interpreted in environmental terms.

(b) Rackham, Sussex

In contrast to Silbury Hill this site is simple and was investigated by one technique only, namely pollen analysis. At the time it was discovered the site was under heathy woodland dominated by young birch. At a depth of 15–20 cm there was a flint flaking floor, apparently of Beaker date. The soil was a leached podzol, characteristic of heathlands today, and a series

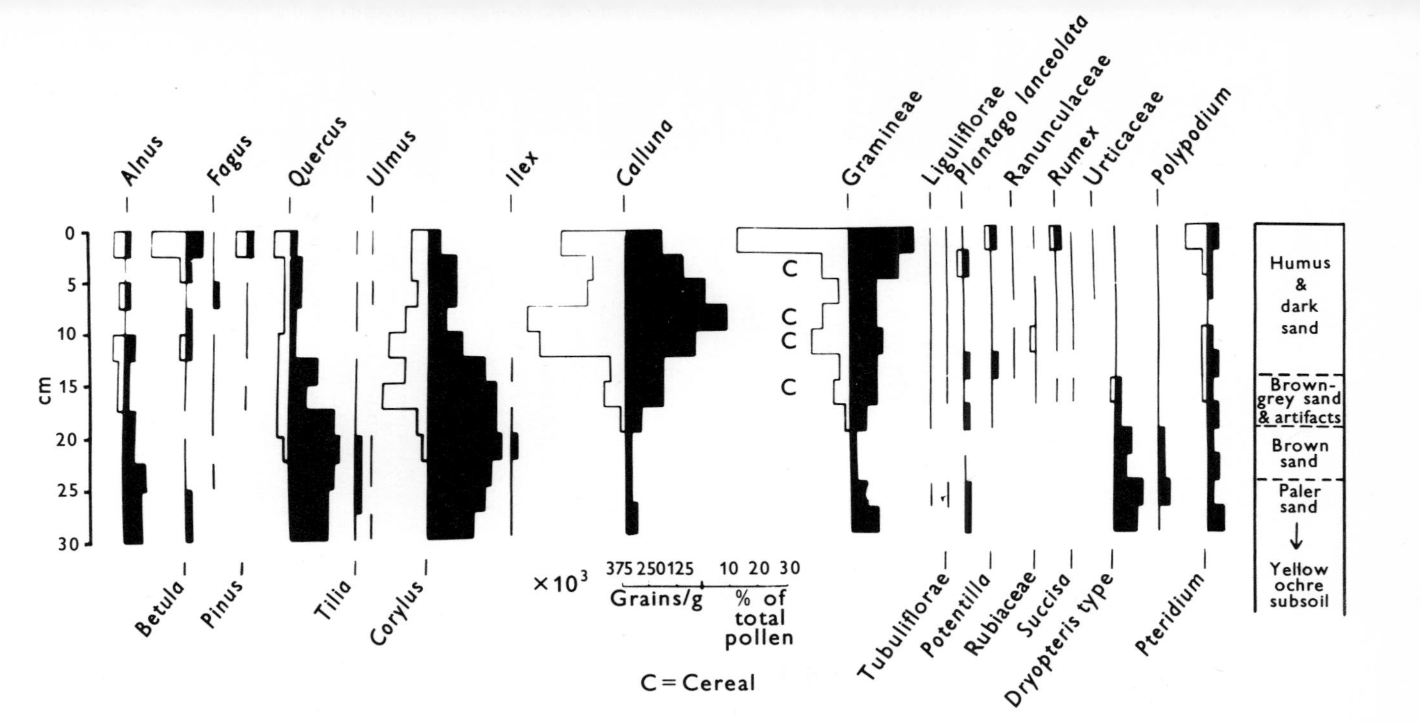

Fig. 6–1 Pollen diagram from a late neolithic site at Rackham, Sussex.

of samples was taken at 2.5 cm intervals throughout the profile for pollen analysis. The resulting analysis is shown in Fig. 6–1.

At the time it was in use the flint flaking floor was presumably at the surface. How did it become buried? The obvious explanation is that dust and sand gradually obliterated it, but this is not borne out by the pollen analysis. When a surface is exposed it receives pollen rain, and, as explained in § 4.4 above, pollen gradually washes down into the soil. If one studies the quantitative distribution of pollen down a soil profile one finds that the greatest concentration of pollen occurs at the surface, and that the amount rapidly falls away with depth (cf. Fig. 4–3). The movement down the profile is slow, for reasons which cannot be discussed here. Consequently, when a soil surface is buried it can be recognized at once in a pollen series by the high concentration of the various pollen types at that level. In this Rackham series (Fig. 6–1) the unshaded histograms represent the pollen quantity or frequency in each sample, and it is apparent that there is no significant increase at the level of the flint flaking floor.

Another explanation is possible. As a result of two or three decades of pollen analyses on heathland sites we know that most of our heathlands have been derived from deciduous forest over the last 4000–5000 years. We also know that the soils under the original forest bore no resemblance to the present acid and infertile heathland soils; indeed, they were cleared for agricultural use. Now this Rackham series has at its base a pollen assemblage representing deciduous forest. Such a forest would have had a biologically active soil in which earthworms were probably present. Let us assume that this flint flaking floor was set in a clearing in the forest (the pollen diagram shows traces of light-demanding species at 25–30 cm). If the site were used and then abandoned, so that the forest returned, all the artifacts would be progressively buried by the action of earthworms in bringing fine material to the surface, thus causing the artifacts to sink uniformly. Darwin showed how Roman tessellated pavements could be buried by this process while still keeping their continuity.

Obviously this process only continues as long as the earthworms are active. The pollen diagram, however, shows a phase of replacement of woodland by heathland, dominated by heather (*Calluna*). This probably happened in later prehistoric times, and is the consequence of forest clearance (see next chapter). Consequently, the soil becomes progressively more acid, earthworms, being intolerant of acidity, disappear, and in due course the soil profile changes from an unbleached brown soil, to the strongly bleached podzol seen today. This one site demonstrates the whole process in a single set of pollen analyses.

7 Human Influences

7.1 The continuity of human influence

In the opening chapter it was shown that man has been an integral component of many ecosystems for a long period of time, and that he had strong ecological tools at his disposal, such as fire, cultivation or the manipulation of grazing. We have now seen that the results of the use of these tools is often detectable from the evidence preserved in archaeological sites.

This may give the archaeologist excavating any given site all that he needs to complete a picture of man, at a given level of culture, set in an environment on which he depends and on which he is having a detectable effect. But archaeologists are also concerned with the changes which go on from millennium to millennium, the development of culture, migration and immigration, invasion and settlement. Similarly, the environmentalist is concerned not just with one episode of man's impact on the environment, but with the effect of the long history of man and ultimately its culmination in the present day. The environment in which we live today is a living carpet which has been trampled on, cared for, abused or neglected by wave upon wave of our predecessors and to some extent each has left an indelible mark. Some features of our landscape can be traced right back to prehistoric actions, such as the destruction of the primary forest which produced the treelessness of our highland zone, which persists today.

In Chapter 2 the contribution of paleoecological studies based on non-archaeological sites was outlined, and now some examples will be given to show how specific sites can produce evidence to fill in the details of this broad canvas.

7.2 The Lake District

The Lake District is a well-circumscribed geographical region and shows evidence of occupation by successive cultures at least from the Mesolithic onwards. When considering this area as an environment for man, the adjacent lowlands and coast should also be taken into consideration; we have increasing evidence that the occupation of the upland areas was not a separate self-contained system but was often linked to lowland areas, perhaps on a seasonal basis.

It was shown in § 2.1 above that stratified deposits such as peats and lake muds are valuable sources of information about past conditions; being so rich in such deposits the Lake District is a ready-made field laboratory for paleoecological studies. A great deal of work has been done in this region, inspired by the late Professor W. H. Pearsall, and carried on by a number of research workers, notably Dr Winifred Pennington. This brief account is based on her book (with Professor Pearsall) *The Lake District*, and other published papers of hers.

Two main strands of evidence have been brought together to reconstruct the landscape history: pollen analysis of lake deposits and peats, and chemical analysis of lake muds. The chemical analyses have also been augumented by analyses of the diatom flora of the muds. Diatoms are microscopic algae which have silica shells; these shells persist and can be identified. They comprise a range of species which flourish in waters of different chemical composition; from the changes in diatom flora through a deposit, therefore, the chemical history of the lake water can be read.

On these grounds it can be shown that after the initial inwash of detritus from melt-water at the end of the last glaciation, a period of stability was reached, apparently associated with the forest cover which developed as the postglacial climate became temperate. This continued until about 5000 years ago, when two new trends began to develop in the lake deposits. First of all there was an increase in the amount of solid matter washed into the lakes; it could also be deduced that the soils of the lake catchments, particularly on the more acidic rocks, were becoming increasingly acidic. Pollen analyses showed that these changes were associated with a reduction of forest cover, the appearance of grasses and weeds and occasionally of cereal pollen. It is apparent that these records are reflecting the impact of neolithic man and the subsequent cultures, all practising agriculture in one form or another. Abundant archaeological remains testify to their presence.

This broad picture is now being filled in in more detail from pollen analysis of tarns in the hills at critical altitudes and adjacent to known areas of occupation, as well as of lowland and coastal sites. It appears that there had been minor clearances made by mesolithic man in the latter half of the fourth millennium B.C. but these were not persistent. When neolithic man arrived with his crops and his animals the deciduous forest reached up to about 750 m. The first noticeable change in the pollen profiles is the so-called 'elm-decline', a phenomenon so widespread that it was assumed that its cause was climatic. Dr Pennington's own work in the Lake District, however, has shown that there was a sudden and actual reduction in the amount of elm pollen being deposited, and it has also been shown that the decline is not coeval everywhere, which makes a climatic explanation less likely. It seems probable that the elm decline was the result of the first inroads of pastoralists into a forested region in which available grazing was very limited. Elm foliage and twigs are much sought after by stock and in the absence of other readily available browse or grazing there would be a concentrated attack on elm.

After this phase clearings were made in the forest and these would provide alternative sources of animal fodder. Charcoal is found in some of the tarn deposits at this stage, showing that fire was being used in the clearance, at any rate along the high-level margins of the forest. It was also shown that maximum clearance was correlated with maximum activity at the factory sites where the characteristic Cumbrian type of stone axe was being made, indicating that deliberate felling was undertaken. The clearances so made were the source of various weed pollens, particularly the ribwort plantain (*Plantago lanceolata*) which previously had occurred in insignificant proportions, but from this stage onwards became a characteristic component of the pollen rain.

It is not possible in this short account to follow the varied ramifications of man's continued influence in this area. We may, however, note some general trends, without implying that the sequences were the same everywhere. It appears that in the uplands and along the Cumbrian coastal plain the forest, once destroyed, never returned. Continued land use, and especially the pressure of grazing animals, prevented the regeneration of the forest. Other places, however, such as the valleys round Rydal Water and Thirlmere, showed successive phases of clearance and regrowth, probably due to shifting cultivation, whilst pasture land was maintained above 200 m.

The forest clearance and the prehistoric agriculture which followed had profound effects on the soils. There was some erosion, indicated by the inwash of mineral matter into tarn deposits, and it could also be shown that the character of the soil humus was changing from the relatively base-rich mild humus to the acidic raw humus, often accompanied by a change from grass cover to heather. Associated with this change, and perhaps the direct result of it at higher altitudes where rainfall is heavy, is the onset of blanket peat formation. The evidence from the Lake District supports the contention that the formation of blanket peat is connected with man's influence on the land (see § 7.4 below).

Though these soil changes were taking place in the hills, there are indications that the steeper slopes remained wooded and the soils retained their fertility. There were also certain fells which were less subject to human pressures and here too the soil fertility did not show the same decline. Geological factors may also play a part in such areas. In some cases, forest clearance did not really become extensive until the Bronze Age, but once it started the familiar pattern was established. Indeed, these changes are inevitable once the primeval forest/soil equilibrium is disrupted; it does not require a change of climate to explain this ecological deterioration.

Nevertheless, if a climatic deterioration does come along, its effect is likely to be more marked on a disturbed environment than on an undisturbed one. There are independent grounds for believing that there was a climatic deterioration round the period 800–500 B.C. when the climate became cooler and wetter. In the lowlands bogs began to grow and timber trackways that were laid down to keep lines of access across the bog were swallowed up by the growing peat even before they showed much signs of wear. At the other extreme, on the hills, the increased growth of blanket peat, and the greater acidification of those soils which were not peat covered, were probably the factor which finally drove the prehistoric farmers off the uplands. In some areas this phase was followed by a regrowth of trees on the hills, showing that it was the activity of man and his animals, and not the state of the soil, that was keeping the forest at bay. Nevertheless, it can be shown that in other places the later stages of prehistoric farming had led to such deterioration of the environment that severe erosion took place from sloping ground, leading to rapid silting of lake basins, and doubtless heavy deposition in river valleys and estuaries.

The environmental work in the Lake District has been carried out almost entirely on stratified deposits, with little reference to the evidence contained in the archaeological sites themselves, though one or two buried land surfaces were analysed for pollen. It is remarkable how much detail has been worked out in this way. However, not all areas are so blessed with strategically placed deposits as the Lake District, and in such cases the archaeological sites themselves can be used to provide similar information. The following examples illustrate this principle in three quite different situations.

7·3 North York Moors

The watershed of the Cleveland Hills was apparently occupied in late mesolithic times, for great numbers of small microliths are emerging from the base of the blanket peat as extensive areas of it erode under today's land use in these areas. It is inconceivable that man could live a hunter-gatherer existence (for this culture existed before agriculture came to Britain) in an environment resembling that of today. A double approach has been made to try to find out what conditions were like in the fourth or fifth millennium B.C. when these people lived there. Both involved the use of pollen analysis. The first was an investigation of the setting in which undisturbed microliths are found. Here it was found that in some cases the flints lay on the surface of the mineral soil, whilst in others they were embedded in the base of the peat. In the former case, charcoal of oak, alder, birch and hazel was found associated with the flints. The pollen analysis showed them to be associated with a forest environment with the same tree species the most abundant. The organic layer immediately above the flints was probably forest humus and not true peat.

In other cases investigated by Dr I. G. Simmons, however, the microliths were lying in peat, and pollen analyses showed evidence of disturbance of the forest such as man could have made by the use of fire. In some profiles a charcoal layer up to 5 cm thick was present in the peat.

The pollen analyses of both these situations demonstrated the increase in light-demanding species associated with human activity. Trees such as birch and hazel showed increased percentages, as did the grasses and heather. Simmons also showed that traces of light-demanding weeds more usually associated with agricultural activities appeared in what seemed to be small clearances associated with this occupation. In particular pollen of cow-wheat (*Melampyrum*) was recognized, a plant typical of open deciduous wood on acid soils.

It is thus made clear that mesolithic man was not in fact living on these hills in a moorland environment but in a forested one—one that would be rich in game, particularly the larger herbivores such as red deer, roe deer, and the huge wild cattle (aurochs). But he was modifying his habitat, and the presence of microliths set in the base of the peat suggests that in the wetter places, perhaps at springheads, peaty conditions were developing. In a later example (§ 7.4) we shall be referring to the possible role of man in the initiation of peat formation, and it may be that even before clearances were made for agriculture the same process was at work. There are grounds for believing, however, that in this case the effect was ephemeral, and on the drier ground, at any rate, the forest restored itself to its original condition.

A study of the ancient soils on Great Ayton Moor on the northern fringe of the North York Moors, gives a clear picture of the sequential impact of man on one small area of land from the Neolithic to the present day. On this piece of moorland, which lies at 300 m altitude on the northern fringe of the Cleveland Hills, there is a remarkable collection of prehistoric monuments: a late neolithic chambered cairn, several round barrows and stone circles of the Bronze Age, and a small rectangular enclosure of the Iron Age. The old land surfaces from each of these three periods were sampled, together with the humus layer of the modern soil. The results are expressed graphically in Fig. 7–1. In neolithic times very little non-tree pollen was recorded, showing the denseness of the forest at this place, but by the Bronze Age the forest had been opened up, allowing the growth of herbs and grasses (non-tree pollen); this opening up of the forest also allowed the growth and flowering of hazel. Cereal pollen was first recorded at this time but the increase in plantain pollen probably indicates that pastoralism was the main land use. By the Iron Age the country was as open as today (high non-tree pollen percentages) and the representation of agricultural weeds was much increased. In all these three prehistoric levels the amount of heather pollen was insignificant, in great contrast to the modern turf which reflects the heather dominance of the moor today. So we see the long history of use of this land, originally under deciduous forest, for agriculture. Comparison of buried soils provides an indication of the increased leaching of the soils through this

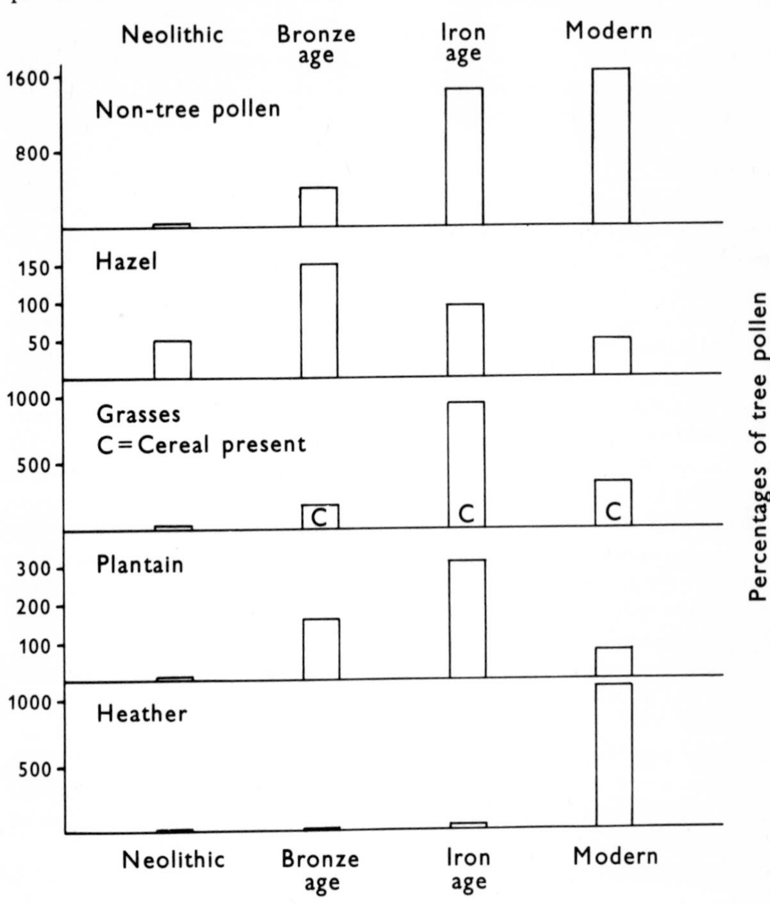

Fig. 7–1 Pollen analysis of buried surfaces at Great Ayton Moor, in the North Riding of Yorkshire.

period, and it was probably this deterioration of the soils which brought about the collapse of agriculture and allowed the land to go over to the acid-tolerant heather. This in turn would lead to the replacement of mild humus (*mull*) by the more acid raw humus (*mor*) and a corresponding decrease of biological activity in the soil. For instance, earthworms would no longer survive. We cannot say just when this change took place; some time after the Iron Age is as far as we can go. It does appear, however, that this long sequence represents the result of man's exploitation of this land over a period of many centuries. As in the Lake District, it is not necessary to invoke any climatic alteration to explain these dramatic changes. Whilst there was a worsening of the climate at about 800–500 B.C., the sequence was already well-advanced by then; climate may have influenced the nature of the ultimate ecosystem, but it was not the causative agent which led to its establishment.

7·4 Goodland, Co. Antrim

In the previous section we studied a sequence on one site that covered some 4000 years from start to finish. Sometimes these changes can be surprisingly rapid. Just inland of Fair Head in Co. Antrim, N. Ireland, lies Goodland Townland. It takes its name from the fertile soil derived from the local exposure of chalk, over which in places lies a thin layer of glacial deposits. A neolithic settlement has been found here, now covered by several feet of blanket peat. Archaeological excavation has revealed some indication of a mesolithic occupation on the site, followed by neolithic forest clearance and cultivation.

Botanical investigations included pollen analysis of the peat and of the underlying buried soil, and seeds were also extracted from the buried soil. In addition, the soil conditions were recorded, backed up by analyses of iron and humus. The soil showed clear evidence of having been sorted by earthworms (see § 5.4), yet it had become acid enough to allow pollen to be preserved (see § 4.4). Here we can again trace clearly the progressive deterioration of the soil following the removal of the tree cover and the exploitation of the soil. In the primeval state the trees would be rooting in the chalk and enriching the soil, and as on Great Ayton Moor the removal of the trees would stop this process and allow the leaching processes in this wet climate to proceed without compensation. Eventually the soil acidity became great enough to eliminate the earthworms and permit the preservation of pollen. The soil profile reflected this change and a thin iron pan soil developed; the iron pan cemented together stones, fragments of pottery and charcoal which had all been buried to the same level by earthworms, thus proving that this thin iron pan soil developed later than the farming phase (Fig. 7–2).

Seed analyses showed that at the surface of the buried soil there was a great preponderance of the seeds of rushes (*Juncus conglomeratus/effusus*), showing that the land surface had become wet. The likely explanation is that this was due to a loss of soil structure associated with the pedological changes already described, so that the sponge-like capacity of the surface of the soil to absorb water was reduced. In this wet climate this would lead to surface waterlogging. The next stage is the beginning of the growth of blanket peat, and analyses through the basal layers of the peat show the replacement of the wetland plants such as *Juncus* by the characteristic plants of the blanket bog.

It seems unlikely that the increased wetness described above was due to a climatic change; the peat started forming at widely different times, apparently related to micro-topography. What was particularly remark-

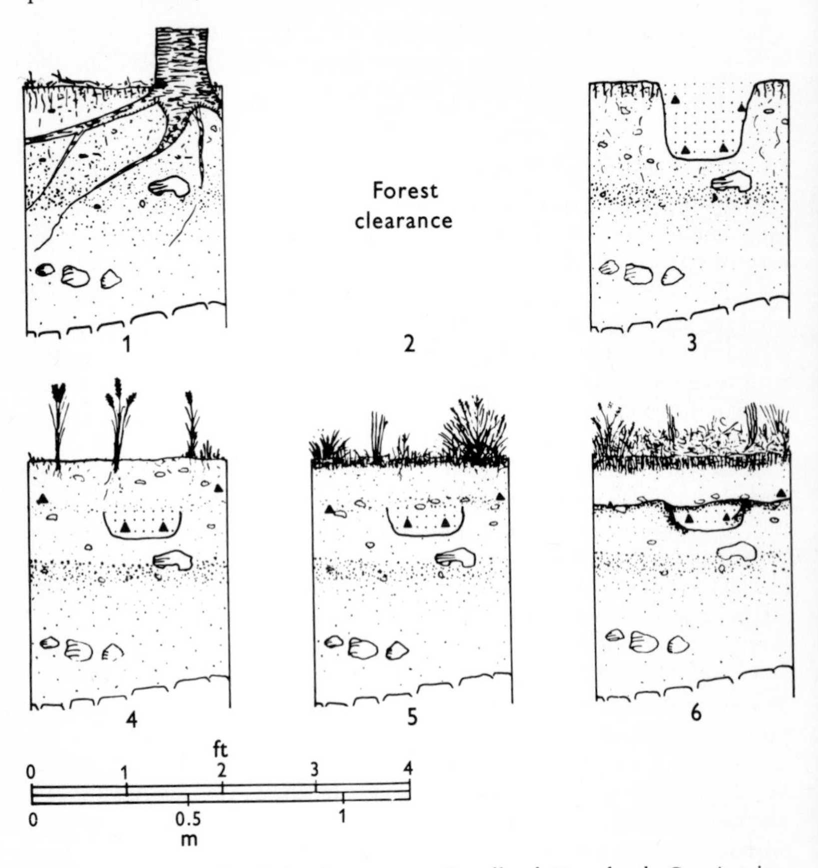

Fig. 7–2 Stages of soil development at Goodland Townland, Co. Antrim. (Courtesy V. B. Proudfoot.)

able was that in some places the base of the peat gave a radio-carbon date of 2200 B.C., showing that this whole sequence in places developed within the neolithic period—a period, incidentally, which does not span a known phase of climatic change.

It had been assumed that the formation of blanket peat was purely a manifestation of climate, but recent pollen work by Dr P. D. Moore has shown that there is usually some evidence of human influence on the landscape in the samples taken at the base of such peat accumulations. He finds that the pollen of plants associated with open conditions or with man-made environments is found in small quantities at these levels. This site of Goodland illustrates one set of processes which could have led to the inception of blanket peat in a wet climate.

7·5 Chalk downs

As a complete contrast to the two foregoing examples, let us turn to the environmental evidence of the interaction of man and his habitat on the chalk of southern England. The chalk gives a highly calcareous, freely draining soil which is therefore not susceptible to the pedological process of acidification and peat formation which have just been described. It produces very fertile soils which were intensively exploited from the neolithic onwards.

Much ecological argument has centred on the question of whether the downs ever carried forest; some have argued that the soil was too free-draining for tree growth, and have seen the floristic character and richness as support for the view that grassland was the climax vegetation. Direct evidence is difficult to obtain, but indirect evidence, particularly from archaeological sites, now suggests that forest did exist, but that it was cleared by the early agriculturalists. Because of the nature of the terrain and its geology, polliniferous deposits are scarce in chalk areas. We now have one or two diagrams, however, which show this early forest stage. One from near Lewes indicates that general forest clearance did not take place till the middle Bronze Age.

Environmental studies of archaeological sites themselves, however, prove that locally, if not over a wider area, clearance took place early in the neolithic. In these calcareous soils the main source of evidence is the molluscan fauna; it can be shown that in a rendzina soil (the characteristic shallow calcareous soil of the chalk, see Fig. 2–1) there is a broad stratification of the mollusc shells, similar to pollen stratification in acid soils, though the cause of this stratification is not fully understood. Such soils also contain features referred to as subsoil hollows: these are lobes of humus-rich or more weathered topsoil which extend down into the more or less unweathered chalk subsoil. They often contain the shells of snails characteristic of a forest habitat, and this suggests that these subsoil hollows are caused by old tree roots; they constitute some of the strongest evidence for the early existence of forest.

Such hollows were found in the paleosol beneath the massive banks at Avebury, and in this soil profile the molluscs showed a change from woodland species at the base to the species of dry grassland in the neolithic ground surface (Fig. 7–3). Similar changes have been found under other neolithic monuments, though the early woodland phase is not always represented so clearly. What is apparent is that by the end of the third millennium B.C. the forest had generally given way to grassland in the vicinity of these neolithic monuments. The molluscs, of course, only have a very local significance and it is not justifiable to make regional deductions from such analyses based on only a few sites.

At this stage the evidence of the molluscs is that clearance was usually for pasture. Parallel pollen analyses have shown that cereals were being

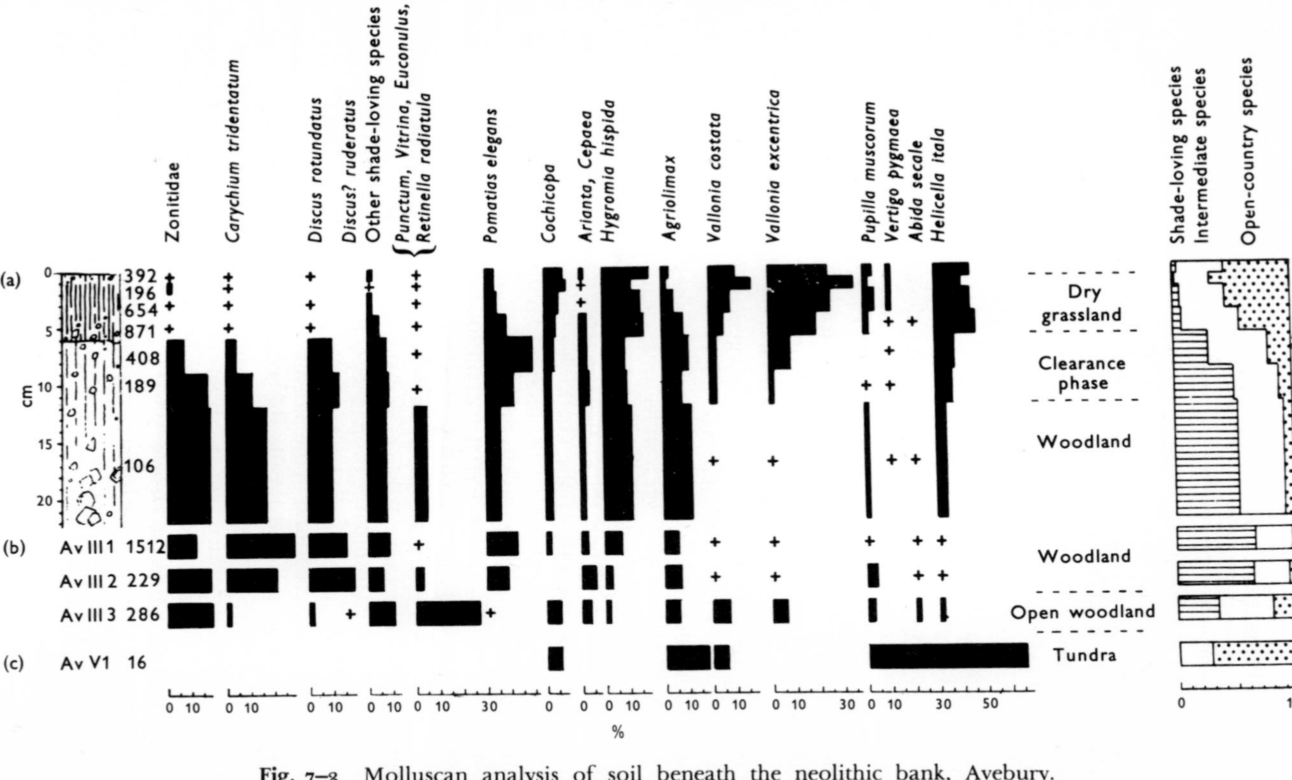

Fig. 7–3 Molluscan analysis of soil beneath the neolithic bank, Avebury. (Courtesy Seminar Press.)

grown, and there is other evidence which suggests that farmyard refuse was being mucked out onto the land even in the neolithic period, which presumably implies arable farming. Under some long barrows (neolithic) plough marks have been found, which bears this out. It may be that the arable use of any plot was only short lived and that it was then turned over to pasture.

It is difficult to establish whether this neolithic impact had any dramatic effect on the soils of the chalk. Recently it has been possible to obtain a radiocarbon date for hillwash on the scarp of the Chiltern Hills. This date, 1960 b.c. (radiocarbon years), suggests that, on areas of steep topography at least, clearance of primary forest did lead to soil erosion. Generally, however, soil movement is not easily demonstrable on the chalk until much later, mid-first millennium onwards. This was associated with the intensification of arable farming which took place in the later prehistoric period and the extension of field systems on to the slopes of the hills. In these situations we often see the outline of the so-called Celtic fields. The lynchets marking the boundary between two fields seem to have built up through progressive soil creep, and mollusc analyses through the bank of a lynchet have been shown to reflect progressive clearance and the establishment of more open conditions.

Individual sites such as Silbury Hill (§ 6.4) can add detail to the general picture, but here on the chalk, as on the acid soils dealt with above, the broad pattern of human influence is clear, and with it the evidence of disruption of the pre-existing ecosystems. The big difference is that whereas site degradation in the acid soil rendered the land almost useless, this was not true in the chalk. There calcareous soils readily develop grassland even after soil removal, and truncated soils can still be used for cereal cultivation; the demand of cereals for nitrogen is not high and seems to have been met adequately even after the turf had been removed.

8 Experiment

8.1 Where experimentation can help

Even with all the corroboration which can be given by co-ordinating the different lines of investigation, one is still aware that interpretation contains an element of uncertainty. It is possible to come to certain conclusions which the facts seem to support, but in making deductions about the distant past there can be no proof that the interpretation is correct. It may be possible, however, to reduce the uncertainty about some of the lines of investigation by subjecting some of the assumptions to experimental testing. It is not possible, of course, to test experimentally processes which take centuries or more to complete, but some of the processes concerned follow an asymptotic course; that is, they are rapid at first and then progressively reach a slow rate of change. In such cases it is possible to obtain information on the rapid trends in the initial stages. Such studies are particularly applicable to decomposition processes, most of which follow this pattern; whilst we may know the course of such changes we are often ignorant about rates.

There are other situations where experiment can give us completely new and sometimes unexpected information; for instance we have no means of knowing the detailed sequence of weathering of a newly-constructed bank or a freshly-cut soil face unless we do an experiment to find out. Archaeologists have used experiments in this sense on a number of occasions, and a good account of them will be found in J. M. Coles' *Archaeology by Experiment*. Some of these experiments have a strong ecological connotation, and one project at least arose from biological interest in the first place. In this chapter we will concentrate on this aspect of environmental work and I shall use the examples of some existing experiments to show the principles. In the Appendix are some suggestions for further experimental work which could usefully be undertaken under the aegis of schools or colleges.

It will be obvious from what was said above that the longer an experiment can run, the more valuable it will be in relation to archaeology. For this reason, continuity of observation and recording needs to be ensured over a period of time. In laboratory work this may mean the handing on of responsibility from one class or team to another, whilst in field work security of tenure of the piece of land on which the experiment is set is of paramount importance. For this reason, such work is more effectively carried out under the aegis of an institution, or a society rather than by individuals.

8.2 The artificial earthworks

At the 1958 meeting of the British Association for the Advancement of Science, archaeologists, anthropologists and biologists combined in a session in recognition of the 150th anniversary of the birth of Charles Darwin. As mentioned earlier (§ 5.4) Darwin drew attention to the archaeological importance of the earthworm, and, indeed, he initiated experiments on the rate of burial of objects by worms which are still the main source of our knowledge about the rates of these processes. Arising from the discussion at this session, plans were laid to institute some new experimental work on as long a time-scale as feasible. In view of the problems of establishing field experiments and servicing them over a long period of time, the final form of the experiments was designed to include both archaeological and biological aspects. Interim reports have been published on these experiments, so I shall confine my observations here to those aspects of direct relevance to this book.

The final plan was to build a linear bank and ditch. The old land surface would be stripped from the site of the ditch and stacked along the centre of the bank to make a turf-core. Turf-cores occur frequently in earthworks and we have already seen that they can provide environmental information.

Within this basic structure objects of archaeological and environmental significance were to be buried, to see how quickly they altered their character and whether they became displaced relative to their original position. The bank and ditch were built to a definite sectional outline and devices were inserted in the bank to record any mass movement. Details will be found in the appropriate publications (e.g. JEWELL, 1963). Finally, it was decided to construct two earthworks of this sort, one on a chalk soil (Overton Down, Wiltshire) and the other on an acid heathland soil (Morden, near Wareham, Dorset).

On completion each bank and ditch was over 30 m long; the bank was 2 m high and had a spread of about 7 m. The ditch was 1.75 m deep, flat bottomed, 2.4 m wide at the bottom, 3 m at the top. The design was to section the bank and ditch at intervals of 2, 4, 8, 16, 32, 64, and (hopefully) 128 years from construction; the positions of the sections were predetermined, and the various buried objects inserted in exactly known positions in the line of each section.

In this short account it is not possible to describe in more than outline the scope of these experiments; at the time of writing both earthworks have been sectioned three times, so in terms of the full span the experiment is in its infancy. We will now list the main factors being investigated and comment briefly on the results to date.

Weathering

Dramatic changes took place in both the bank and the ditch, under the

Fig. 8–1 The Overton Down earthwork, (**a**) on completion, and (**b**) after 3 winters' weathering. (Courtesy The Prehistoric Society.)

effect of weathering (Fig. 8–1). In both earthworks the ditch weathered away below the soil, eventually undercutting the soil so that great swathes of topsoil fell into the bottom of the ditch. The implication of this from the point of view of both archaeological and environmental interpretation is profound. The bank surface of the Overton Down earthwork was composed of large chalk lumps which broke down to a fine gravel, apparently due mainly to wetting and drying. The Wareham bank, however, was sandy and this developed a gully system and was also subject to wind erosion. Both banks showed settlement due to the compaction of the turf core. An unexpected observation was that the two sides of each ditch weathered at almost identical rates despite the different orientation relative to prevailing weather and the sun.

Pollen movement

Both earthworks included an experiment designed to quantify the rate of pollen movement through soils beneath a bank. Before the bank was thrown up the ground surface was dusted with *Lycopodium* powder; these spores are very resistant to decomposition, and, being alien to the local flora, are easily located in pollen samples. At the time of each sectioning a series of samples was taken through the buried soil and up into the bank. The results were dramatic: at Wareham the *Lycopodium* spores remained within the old land surface; there was no detectable downwash in 9 years. At Overton Down not only could the spores be traced some 10 cm or more down into the buried soil, but measurable quantities were found to have moved *upwards* into the bank (Table 1). This is clearly the result of

Table 1 The distribution of *Lycopodium* spores beneath the Overton Down bank after two (1962) and four (1964) years: Series 1 Beneath the turf core; Series 2 Under chalk rubble near centre of bank; Series 3 Under chalk rubble near edge of bank.

Depth (cm)	*Series 1*		*Series 2*		*Series 3*	
	1962	*1964*	*1962*	*1964*	*1962*	*1964*
7.5–5.0	—	18.5	—	—	—	—
5.0–2.5	0.4	9.1	1.5	5.5	3.0	20.2
2.5–0	11.3	34.6	5.9	21.4	14.5	24.0
0–2.5	87.7	33.8	63.3	39.9	69.0	36.6
2.5–5.0	0.7	3.6	22.6	15.1	11.5	11.3
5.0–7.5	—	0.1	5.1	14.0	2.0	5.4
7.5–10.0	—	0.1	1.7	2.8	0.0	2.0
10.0–12.5	—	0.2	0.0	1.0	—	0.1
12.5–15.0	—	0.0	—	0.3	—	0.1
15.0–17.5	—	0.0	—	0.0	—	0.0
17.5–20.0	—	—	—	—	—	0.1

earthworm activity; the possibility of such upward movement had not previously been appreciated.

Microbiological activity

Space does not permit more than an outline of the observations made on the organic materials buried in the earthworks. These materials were buried in two contexts, one underneath the turf cover and therefore in an organic environment, and the other in the mineral body of the bank. The following objects were buried: charred and uncharred billets of oak and hazel; cooked and uncooked animal bones; cotton, linen and wool cloths; leather; ropes of hemp and flax fibre; and human bone for blood group tests.

It was possible to trace the progressive deterioration of these materials in both earthworks; in some cases decay was very rapid (e.g. cotton cloth) and recovery was difficult. Generally speaking there was little difference in the early years between the organic and mineral settings, though an unexpected difference developed with the wood samples. In the chalk bank there was little difference, but in the bank on heathland soil the billets rotted very rapidly in the turf context, being virtually irrecoverable after 9 years. The implications of this need further investigation.

Colonization of bank and ditch

Botanical records of the spread of vegetation to the bank and on into the ditch were kept. One of the earliest observations was that a continuous layer of blue-green algae, which are potential nitrogen fixers, developed under the surface of the bank. Intermittent colonization by species with wind-carried seeds was the next stage, but did not persist. Eventually the lower levels of the bank became well vegetated, particularly on the leeward side. The ditch filling also became grassed over once active undercutting had ceased. The plants on turves which fell into the ditch continued to grow (unless they were buried) and formed a well-established plant cover.

8.3 An experimental farm

A Research Committee of the British Association for the Advancement of Science was set up to investigate ancient fields. It found that the interpretation of old field systems often involved assumptions some of which could be checked by experiment. To shed light on a variety of such problems, an experimental farm, using only Iron Age crops, animal breeds (as far as possible) and techniques, has been established at Butser Hill, near Petersfield, Hants. This is now in its third year, and has already provided information about crop husbandry, weeds, soil effects, animal health and so on. Again some interesting and unexpected results have emerged. For instance, we now have actual figures for the yield of various primitive cereals, and it has also been shown that cereals such as emmer wheat may have a protein content of 19–20%, compared with the 11% of modern bread wheats. In due course more long-term observations will be made; for example, the rate of cultivation should help to answer some outstanding archaeological questions, such as why and how quickly lynchets develop.

If finance is available to keep the experiment running long enough, valuable data on the rate of burial of objects by earthworms should emerge. Already the effects of shallow or scratch ploughing on earthworm populations can be compared with the effects of modern mould-board ploughing. This experiment is manned by a permanent staff and it has only been possible here to pick out one or two of the many lines of enquiry which are in hand.

Appendix

A. Experiments on decay rates

It has been shown that the extraction of evidence from archaeological sites depends very much upon the conditions of preservation of organic remains. It is known that different materials decay at different rates under identical conditions, and that the same materials decay at different rates under different environmental conditions, but little precise knowledge is available on these questions.

In suggesting projects for experimental work in schools, outlines only are given. In most cases there has been no comparable work, and standard methods have not been devised. In any such work the value of replication should be appreciated; this, of course, requires duplication of techniques, and should therefore be suitable for classwork.

(a) Insect remains

By burying the bodies of different species of insects in the soil and following the course of decay, comparative data on differential decomposition can be obtained. For this purpose, the insects chosen should be species which have strong ecological affinities and which are reasonably common. Beetles are the most suitable, but data for other groups would be valuable.

The most useful approach is to bury individual insects of known dry weight and retrieve them at fixed time intervals. A period of up to six months is probably necessary. After retrieval their dry weight should be determined. It is desirable that the soil should remain as undisturbed as possible, so no attempt should be made to re-inter extracted specimens.

One of the problems arising here is that the insect bodies are likely to be dispersed by soil fauna. They should therefore be enclosed in fine nylon bags of such a mesh that only micro-organisms can get at them. It would be useful to know how much they are attacked by the larger fauna, but if they cannot be retrieved the experiment breaks down. This experiment could usefully be carried out on a variety of soils for comparative purposes.

A limited experiment on these lines has been carried out by Dr Martin Speight at the Institute of Archaeology, London, but a great deal more needs to be done.

(b) Pollen

Dr A. J. Havinga, of the Netherlands, has carried out experiments on the decay of pollen of different species in different micro-environments (see BROOKS *et al.* eds., 1971). His results are striking, but do not entirely accord with field experience. His technique was to mix pollen of known species into an inert filling of clay or sand and bury them in different environments. They can then be extracted at intervals and the state of the pollen estimated. Other techniques could probably usefully be tried. This experiment requires the use of a good microscope, suitable chemical facilities (e.g. for acetolysis) and some knowledge of pollen form and structure, but with experienced guidance could produce valuable additions to our at present incomplete and ambiguous knowledge of a very important field.

(c) Wood

Wood billets were included in the experimental earthworks (see section 8.2) but there are many other contexts in which experiments could be constructed. In the experimental earthworks (see JEWELL, 1963) only two species were used; others could be tried, but remember that many species present in Britain today have been introduced in the last few centuries. It is sometimes argued that charring preserves the wood longer, a point also tested in the earthworks but one on which more data would be useful.

A good deal of information already exists on the decay of wood in the field, but further work would be useful on species of no commercial value and on preservation in habitats not normally found on land where timber is used; for instance, swamp or bog conditions.

Though detailed microscopic investigation of the wood, especially if combined with mycological investigation, would be required for a fully scientific investigation, a simple loss of density analysis could give an indication of the overall rate of decay of the wood. This should, of course, be based on dry weight.

This experiment is likely to take several years before the results are of interest, and the experience from the experimental earthworks would suggest that good replication is desirable.

(d) Textiles

Techniques have been devised for measuring the strength of fibres, and this principle could be applied to assess the rate of decay of various types of fibre such as cellulose (cotton), lignified fibre (flax), or protein fibre (wool). By burying standard strands of material for different lengths of time, and testing their breaking strength at intervals, useful comparative results could be obtained which would relate to the preservation of the fibres, and therefore of textiles made from them, in different conditions of burial.

A. 2 Earthworm activity

The simple experiments carried out by Darwin, in which he deposited stones, pieces of brick, and so on, on the ground and left them to be buried by worms, are well worth repeating. It will take perhaps up to 20 years before really dramatic effects are achieved, but such an experiment could form part of the on-going programme of a school archaeological society and could be set up in a corner of the school playing field. A sampling pattern over a number of years would have to be worked out, but the course of the experiment could probably be followed by some form of fine probe that would locate the buried objects without disturbing them.

A short-term experiment involving earthworms is the assessment of their effect on soil pollen. This can be done by keeping earthworms in flower-pots or other suitable containers, and adding to the surface of the soil a sprinkling of some pollen type or spore (e.g. *Lycopodium* powder) which is alien to the soil. The worms will feed through the soil, which should be humus rich, and their feeding disperses the pollen through the soil mass. The pots should be watered by a wick from below so that downwash of pollen is eliminated.

At the end of the experiment the soil is sampled serially through the pot and the pollen extracted by the standard techniques used in pollen analysis.

A. 3 Land snails

Serial sampling of modern grassland soils in calcareous areas may produce evidence of local short-term ecological change. Dating such changes may not be easy, unless documentary sources referring to land-use changes are available.

Experiments similar to those described in section A.2 above can be done using snail shells rather than pollen. The results of such experiments could influence the way in which molluscan faunas are interpreted, and may also cast light on how successive assemblages become stratified in soils.

Further Reading

ASHBY, M. (1969). *An Introduction to Plant Ecology* (2nd. edn.). Macmillan, London.

ATKINSON, R. J. C. (1957). Worms and Weathering. *Antiquity,* **31**, 219.

BRAY, W. and TRUMP, D. (1972). *Dictionary of Archaeology*. Penguin Books, London.

BROOKS, J., GRANT, P. R., MUIR, M. D., VAN GIJZEL, P. and SHAW, G. (Eds.). (1971) *Sporopollenin*. Academic Press, London.

COLES, J. (1972). *Field Archaeology in Britain*. Methuen, London.

COLES, J. (1973). *Archaeology by Experiment*. Hutchinson University Library, London.

COOPE, G. R. (1970). Interpretations of Quaternary Insect Fossils. *A. Rev. Ent.* **15**, 97.

DARWIN, C. (1966). *Darwin on Humus and the Earthworm*. (New edn.) Faber, London.

DIMBLEBY, G. W. (1967). *Plants and Archaeology*. John Baker, London.

EVANS, J. G. (1972). *Land Snails in Archeaology*. Seminar Press, London.

EVANS, J. G. (1975). *The Environment of Early Man in the British Isles*. Elek, London.

JEWELL, P. A. (Ed.). (1963). *The Experimental Earthwork on Overton Down, Wiltshire, 1960*. British Association for the Advancement of Science, London.

LIMBREY, S. (1975). *Soils and Archaeology*. Academic Press, London.

MACFADYEN, A. (1963). *Animal Ecology. Aims and Methods*. Pitman, London.

PEARSALL, W. H. and PENNINGTON, W. (1973). *The Lake District*. Collins New Naturalist, London.

PENNINGTON, W. (1974). *History of the British Vegetation* (2nd. edn.). English University Press, London.

PLACE, R. (1968). *Introduction to Archaeology*. Newnes, London.

Acknowledgments

This book would not exist without the interest and encouragement of my friend and mentor Margaret Levi. In the wake of the 1999 "Battle in Seattle," where environmental activists wearing turtle costumes united with hard-hatted unionists, larger-than-life puppets theatrically drew attention to global injustices, Gore-Tex–clad locals came out in droves, and the WTO Ministerial Conference was effectively shut down, Margaret instigated the University of Washington's WTO History Project.* As her team—in which I assumed a small role—interviewed activists; collected documents, maps, costumes, and ephemera; and collaborated with amazing librarians, some of us began thinking about how to study "global citizenship" as it is embedded in social movements that mobilize the resources of activists in the global North to abet the efforts of activists in the global South. We focused on campaigns that were strong in our locale, notably Fair Trade coffee. The rest is history, a bit of which I recount in this volume's introduction.

I did not research this book alone. I am immensely grateful to Renee Chou, Rebecca Kahn, Kira Luna, Louise Ly, Marie Murphy, Rebecca Jo

* WTO History Project (http://depts.washington.edu/wtohist), accessed September 7, 2011.

Too Smart for Trouble

Sharon Scott
With Nicholas, the Cocker Spaniel

Illustrated by George Phillips

Published by
Human Resource Development Press, Inc.
22 Amherst Road, Amherst, Massachusetts 01002
(413) 253-3488 (Mass.)
1-800-822-2801 (outside Mass.)

First Printing, January, 1990
Second Printing, August,1990
ISBN 0-87425-121-4

To the Little Ones
who I hope will lead the way
to a more sensitive, kind world.

About the Authors

Sharon Scott is a Licensed Professional Counselor who has trained over 250,000 youth and adults internationally in her proven effective Peer Pressure Reversal techniques. Her Dallas-based consulting firm offers training to school districts, social service agencies, religious groups, and parent and youth groups. With a master's degree in Human Relations and Community Affairs, she has developed many innovative programs to combat negative peer pressure. She is former director of the nationally recognized First Offender Program of the Dallas Police Department.

Ms. Scott has authored four widely acclaimed books: ***Peer Pressure Reversal*** and ***How to Say No and Keep Your Friends,*** about managing negative peer pressure; and ***Positive Peer Groups*** and ***When to Say Yes!***, about encouraging positive peer pressure. She writes a "Positive Parenting" column for numerous newsletters. She is also a frequent local and national radio and television talk show guest, including appearances on "Good Morning Australia," "Hour Magazine," and CNN's "Sonja Live from L.A."

She is a featured speaker at numerous conferences, including PRIDE '86, '87, '88, and '89 International Conference on Drugs, International Committee for the Prevention of Alcoholism and Drug Dependency, UNESCO Southeast Asian Conference on Drug Abuse in Schools, National Youth for Christ, Texas Police Association, American Association for Counseling and Development, and PRIDE Canada '87, '88, and '89.

Nicholas, Ms. Scott's "co-author," is a four-year-old cocker spaniel. His degree is in dog obedience from Richland College. In his spare time, he does volunteer

work for the Dallas Society for the Prevention of Cruelty to Animals, visiting nursing homes and schools. Every year at the State Fair of Texas, Nicholas "mans" the SPCA's humane education booth and gives away nose kisses for donations to help his homeless animal friends.

Preface

Today's children are growing up faster than any other generation. They are being asked to make adult decisions in their preteens and often even younger. If we are going to slow down the tragedies that have been climbing yearly with our children—crime, drug use, pregnancy, dropout rate, and suicide—then we must teach prevention strategies at very young ages. We must realize that when our children are asked to do things wrong, it will not be by a stranger on the corner but by their friends that they spend the most time with. Negative peer pressure is subtle, and we must prepare them for it. Just knowing right from wrong and to say no is not sufficient to enable the child to deal with the intensity of modern-day peer pressure.

Little children must *daily* decide how to manage negative peer pressure. They may be dared to do pranks, fight someone, or smoke cigarettes. They may be teased to call people names or ride their bikes too far from home. Poor decisions in these areas can cause them to get in trouble, get hurt, or even die.

Too Smart for Trouble has been written as a read-along guide to instruct children *how* to say no to trouble, yet still be liked. It is far more effective and less stressful for all if we teach children how to deal with trouble invitations *before* they happen, rather than punish or treat the problem after it has occurred.

I want to thank the following children who helped me "edit" this book for clarity and interest, and helped decide whether the words were "too big": Adam McCane, grade 3; Beth Rosamond, grade 3; Andrew Strickland, grade 1; Amanda Muecke, grade 3; David Carver, grade 1; and Sarah Plunk, grade 1. I also appreciate the efficiency

(and laughter) contributed by my office manager, Michelle Cooper, who so carefully typed the manuscript. I also want to thank my publisher and the staff of Human Resource Development Press, who continue to be supportive and helpful. And a special thank-you to George Phillips for his beautiful drawings that bring Nicholas to life on these pages.

Dallas, Texas
July 1989

Sharon Scott

Beginning Note to Parent/Teacher:

Some suggestions that will help the book be most effective for your child or student:

1. Read it first yourself.
2. Then read it to or with the child. Realize that young children's comprehension varies greatly, so there may be points that you need to elaborate on to help the child understand. An example would be in Chapter One. Most young children want to blame the other children for Tim's accident. It is important that they learn that Tim's decision allowed the accident to happen. No one made him do wrong—just as your child won't be made to do wrong by peers. They may be asked or teased, but not usually made to do wrong. If the child does not grasp this concept, you may want to give other examples.
3. Read the book together in small amounts—no more than one chapter at a time. Then have some light discussion about what was read. You may have them tell what they learned to the other parent, or to their doll or dog.
4. Avoid lecturing while reading (e.g., "I hope you are listening to this. You need it." Or, "Remember when you made that stupid decision with your friend? This book may help you keep from doing that again.") Do not use the book to chastise poor decision making. It should always be used on a positive note.

5. The child may want to color the pictures after you have read a chapter.
6. In Chapter 5, "Let's Practice!", let the child answer (or act out) what he would say or do to the bad peer pressure *before* reading the next page about how the child in the story handled it. This gives your child the opportunity to think and practice.
7. Read with enthusiasm and with emotion in your voice—especially on the stories related by Nicholas, the dog who serves as "co-author." Both of you should have fun as you learn!

TABLE OF CONTENTS

Chapter 1

Introduction to Making Decisions

This book is about making decisions—*good* decisions. What is a decision? It is a choice. You make decisions every day—like what time to get up, what you are going to wear, or what you are going to eat. Those decisions are easy most of the time.

Sometimes decisions are harder—like who to pick for friends, how you will spend the $5 you got for your birthday, or when to do your chores.

When you were very young, your parents probably made all your decisions for you. Now that you are getting older and doing more things on your own, it is time to learn more about how to make good decisions.

As you get older, you will spend a lot of time making some decisions, such as what you are going to be when you grow up, or where you are going to college. Your parents have a lot of decisions to make, too. They have to decide where to live, how much money they have to buy a car, and whether to take a family vacation. Everybody, but tiny babies, has decisions to make.

Wait just a minute. Nicholas is saying something.

My dog, Nicholas, says dogs have to make decisions, too. In fact, he has asked to help me write this book for

you. Since he is a good and smart dog, I think I will let him help.

Let me introduce you to him. "Nicholas, these are the boys and girls who want to learn how to make good decisions. And, kids, this is Nicholas." He is a four-year-old blonde cocker spaniel. He has two brothers and one sister: Shawn, another blonde cocker spaniel; Cedric, an orange cat; and Mandy, a black and white spotted terrier. Nicholas has been to college. He took dog obedience classes at Richland College in Dallas, Texas.

Nicholas may have a lot of things to say. We will put his picture and paw prints around anything he says.

Nicholas with Cedric, Mandy, and Shawn

Decision making can be fun. It helps you to have fun with friends, please your parents and teachers, and stay out of trouble!

This book is about a special kind of decision, called a peer pressure decision. A peer is someone close to your age. Pressure is words people may say to try to get you to do what they want you to do.

There is both good and bad peer pressure.

Which of these is your peer?

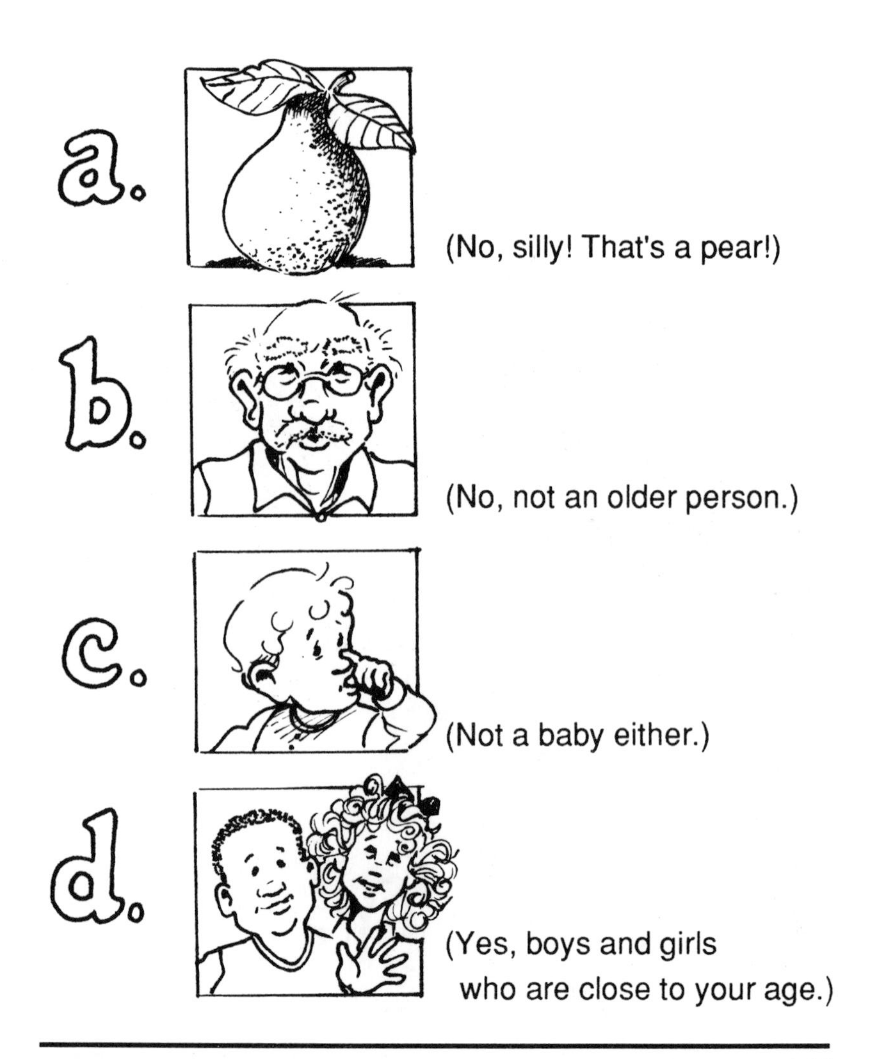

a. (No, silly! That's a pear!)

b. (No, not an older person.)

c. (Not a baby either.)

d. (Yes, boys and girls who are close to your age.)

Good peer pressure is when someone tries to talk you into doing something good.

Have you ever heard someone say:

"Mary, if we both study hard we can make a good grade tomorrow at school." Or,

"Run, Sam, make that touchdown; yea!" Or,

"Tina, don't ride your bike there. It's dangerous."

Those are examples of good peer pressure. Good peer pressure can make you happy.

Nicholas says, "I've had good peer pressure before. One day, when I was still a puppy, I was playing in the park and started to run fast through the trees. Shawn, my big brother, warned me that there was a deep creek at the bottom of the hill. He helped me make a good decision to look carefully where I run."

Bad peer pressure can cause you to get hurt or in trouble. Bad peer pressure might sound like this:

> *"Can I copy your homework? I'll like you if you do."* Or,
>
> *"Let's throw rocks at cars. We won't get caught."*
>
> Or,
>
> *"Have a cigarette. You're a chicken if you don't smoke."*

Bad peer pressure is no fun at all.

Nicholas has had lots of bad peer pressure. He says, "One time a neighbor dog tried to talk me into chasing cats and was telling me that we would not get caught. I thought that was mean.

"Another time while playing in the back yard, a friend and I found some open beer bottles. Someone had thrown them over the fence. They still had some beer left in them. My friend was excited and tried to talk me into trying it. He said some dogs on television even sell beer. He told me that I was acting like a baby puppy when I refused to drink it. I also told him I would never do a beer commercial.

"Sometimes it is hard to say no to bad peer pressure, but we must learn how. Then we can stay out of trouble. And, if I stay out of trouble everybody is happier, including me!"

Things that kids may say to you to try to get you to do something wrong with them are:

"You're a baby if you don't do this."

"Are you chicken?"

"If you were my friend, you would do this with me."

"We won't get caught."

"Be cool."

"You're a real wimp."

"What a nerd!"

"I dare you to do this."

"Nobody will find out."

"Don't act like a sissy."

"Mama's baby."

"Come on, everyone is doing it."

"I'm not going to invite you to my birthday party if you don't do this with me."

"It's no big deal."

I'm sure you've heard some of these. If you do not know how to handle put-downs and teasing, you can easily get into trouble.

Usually friends do not *make* us do things wrong. They just beg us, or tease us, or put us down.

The school was on a big, busy street. Tim, a third grader, usually walked home after school. He knew he should use the crosswalk to get across the street safely. One day, some older students started teasing him and his friends about using the crosswalk. The big kids called Tim a "baby" and showed him their shortcut to go home.

But the shortcut was farther down the street. There was no crosswalk or crossing guard there. The next day when the teasing started again, Tim tried to prove to the big kids that he was not scared. He crossed the street where they had crossed. Tim got hit by a car. He got hurt badly. He broke both legs and an arm. He stayed in the hospital a long time. He missed so much school that he got far behind in his school work and did not pass to the next grade.

Whose fault was it that Tim got hurt? You may think it was the big kids' fault. But did they *make* him take the shortcut? No! The big kids just used words on him. It was Tim's own fault that he got hurt. The older kids were unkind, but they didn't *make* him do wrong.

Tim made his own bad decision. He did not know how to handle bad peer pressure.

Sometimes it can be hard to say no when friends ask you to do something stupid or wrong. You are afraid they will not like you if you do not do what they ask. But friends do not stay mad at you for long. A few hours or maybe a day is the longest time they stay mad. Big deal! You can live with that.

If you learn to make good decisions, then you are in control of yourself. You know what is right and what is wrong for you. None of your friends your age should be making decisions for you.

You will not lose *real* friends by making good decisions for yourself.

Chapter 2

Stories About Bad Peer Pressure

The story about Tim was true. Let's look at some other true examples of what can happen when you make poor decisions while you're with peers. Remember what happened to poor Tim? He ended up badly hurt because he did not know what to say to bad peer pressure. Bad peer pressure can get you hurt, arrested, or in big trouble. These stories will show you examples of what can happen.

Bill's Story

Bill was a fourth grader who wanted to be an astronaut. He built model rockets with his father. When he got older, he even planned to go to space camp.

School grades were very important to him. He always got the highest grades in his class. He always turned in his work on time. Sometimes he even helped the teacher.

Some of the students in his class were jealous of his grades. They made fun of him and called him the teacher's pet. Bill did not like to be called names. He was not sure what to say to this bad peer pressure.

Bill started doing poor work to try to get the teasing to stop. His next report card was lower than it had ever been. Bill was sad.

Bill did not know what to say to bad peer pressure. He needs to learn that he can make good grades and have good friends too.

Ann's Story

Ann had her friend, Jessica, over to play on Saturday morning. They watched cartoons and played some games.

Jessica told Ann that she did not like the new girl at school named Robin. Jessica said, "Ann, if you are my best friend, then you won't talk to Robin."

Ann was not sure what to do. She really liked the new girl. She also liked Jessica and did not want to lose Jessica as a friend. Finally, Ann said, "Okay, I won't talk to Robin anymore."

Ann did not know what to say to bad peer pressure. She needs to learn that a real friend will let you talk to other people, too. You can have more than one friend.

Ted's Story

Ted and his friend, Phil, had stopped at the store to buy a soft drink. They had been riding bicycles and were hot and thirsty.

Suddenly Phil whispered to Ted, "Let's take some candy."

At first, Ted said, "No, that's wrong."

Then Phil said, "We won't get caught. You're a chicken if you don't take some candy."

Ted forgot to think how wrong stealing was. And he did not like being called names. He grabbed some candy and put it in his pocket. When he started to leave the store, the manager stopped both boys and asked what they had in their pockets.

The manager called the police. The police came and talked to the boys. Then the police called their parents. Boy, were they in trouble!

Ted did not know what to say to bad peer pressure. He needs to learn that friends sometimes have stupid ideas. He needs to remember that friends do not *make* you do things wrong. But sometimes they may use words to tease you. Ted needs to learn what to say to bad peer pressure!

Maria's Story

Maria had just walked into class. Her friend, Tina, came up to her and said hi. Then Tina whispered, "I didn't finish my spelling paper. Did you finish your paper?"

Maria said that she had finished it.

Next Tina said, "Then can I see some of your answers? Please! Just this one time!"

Maria said, "No, that's cheating."

But Tina kept on talking to Maria. She begged her for her answers. She finally said, "Everybody cheats. And if you were my real friend, then you would give me your answers." Maria finally gave Tina her answers.

What a bad decision! Everybody does not cheat. Maria forgot to remember that. The teacher caught the girls cheating. They both made a zero on their papers. They were sent to the principal's office. Their parents were called. Both girls got in a lot of trouble. And their parents were disappointed in them.

Maria did not stand up for herself. She did not know what to say to bad peer pressure.

Nicholas's Story

Nicholas has a bad peer pressure story that he wants to tell. Nicholas says, "I was in my back yard playing with Shawn and Mandy. Some workmen had been working in the yard. When the men left, they forgot to close the gate.

"Shawn and Mandy were excited to see the gate open. They yelled to me to come outside the fence and play with them. I knew that I was supposed to stay in the yard. But they said it would be fun. They even dared me. I finally went out to play with them in the alley. Suddenly a boy on a motorcycle came around a corner. The boy was going too fast and hit Shawn. Shawn was bleeding all over his head. He had to go to the hospital and get stitches and a shot."

Nicholas did not know what to say to bad peer pressure. He listened to friends. He should have listened to *himself.* He knows right from wrong.

Chapter 3

Five Kids and the Creek

This is a make-believe story about bad peer pressure. Let's make up silly names for five kids. We will pretend there are two boys we will call Walter Wimp and Macho Mike. The three girls we will call Sylvia Snob, Curious Carol, and Christine Confident.

These five kids live on the same street. One Saturday they were all outside playing. They are supposed to play only on their block.

A bigger kid they all liked rode by on his bicycle. He told them to hurry down to the creek. He said there was something in the water to see. He also said that all the older kids were riding down there. He then said, "If you all are not babies, then come on to the creek."

Let's look at what these five kids said to the big kid.

Walter Wimp
Christine Confident
Sylvia Snob
Macho Mike
Curious Carol

Sylvia Snob stuck her nose up in the air. She walked away saying, "You are so stupid. And you're ugly, too. I'm not going."

How did Sylvia Snob do in knowing what to say to the bad peer pressure? She said no to the trouble. That was good. But she acted mean and stuck up. That was bad. She probably lost a friend.

What did Walter Wimp say to the big boy? He looked at the ground and in a soft voice said, "I don't think I should go down there. We might get caught." The big boy kept on teasing Walter. Walter finally gave in and rode down to the creek.

How did Walter Wimp do? Not very good. He was too weak. He did not sound like he meant it when he first said no. Walter will probably get in trouble.

Let's see what Macho Mike said to the big kid. "Well, if everyone is going to be there, I'll be there too! I am sure everyone will want to see me."

How did Macho Mike do? Lousy. He is afraid to be left out. He is too cool. He shows off too much. And sometimes Mike talks himself into trouble!

When Curious Carol was asked to go to the creek, she quickly said yes. She is so nosey that she does what anyone asks. Curious Carol did not know what to say to the bad peer pressure either. She forgot to think before she gave her answer.

None of these four kids knew what to say to bad peer pressure! Let's see if Christine Confident can do better than they did.

When Christine was asked by the big kid to go to the creek, she knew that it would upset her parents. She and her friends were supposed to play only on their block. She also remembered that the creek was deep, and someone might get hurt.

So she said to the big kid, "No, we could get hurt or in trouble."

When the big kid started to tease her, she said, "No way! I'm going to ride down to Jane's house and see what she's doing. Bye." And off she went with a *better* idea.

Sylvia Snob, Walter Wimp, Macho Mike, and Curious Carol did not know what to say to bad peer pressure. But Christine Confident did! She uses her brain to think. Christine is smart. She's in control.

Do you want to be like Sylvia Snob? Or Walter Wimp? How about Macho Mike or Curious Carol? No! I bet you want to be more like Christine Confident. Boys and girls that know what to say to bad peer pressure, like Christine, have fun with their friends. And they stay out of trouble too!

"What did you say, Nicholas?"

"You want to know what was in the creek."

Nothing! The big kid was just trying to show off.

Sometimes friends want us to do things that sound like fun. But we must still learn how to decide if it is trouble.

Chapter 4

Too Smart for Trouble

Let's learn what to say to bad peer pressure. You can learn how to be more like Christine Confident.

You will learn to be ***Too Smart for Trouble.*** It has three parts to it. ***Too Smart for Trouble*** is easy and fun to learn. ***Too Smart for Trouble*** will show you what to say to bad peer pressure.

The three parts to ***Too Smart for Trouble*** are:

1. Look and Listen

2. Think: Is It Good or Bad?

3. Say No to Trouble

Let's learn what each of these parts means.

Part 1: Look and Listen

This first part tells you to use your eyes and ears to see what is going on around you. You need to become a "trouble policeman." A policeman looks for clues to trouble. So should you.

Look at your friends. See if they are acting nervous. Are they huddled in a group like they have a secret? See if they are acting sneaky or whispering. Are they acting tough or trying to show off?

Look at where they want you to go. Is it dark, or could it be dangerous? Is it someplace that you are not supposed to go?

If you see any of these clues, then your friend may be planning to do something wrong. And they may want you to be with them.

You should also listen carefully to what your friends are saying. Are they begging you? Are they calling you names? Or are they daring you? Do their voices sound mean?

If you hear any of these clues, then your friends may be planning to do something wrong. And they may want you to do something wrong with them.

"I know how to look and listen!", says Nicholas. "I always watch for clues to trouble. When Cedric, the cat, tries to sneak food off the counter, he starts acting sneaky. He also looks around to see if anyone is watching. Once he whispered for me to watch to see if Mom was coming. I knew he was up to no good!"

Part 2: Think: Is It Good or Bad?

If you see or hear clues to trouble, you have to think if it is good or bad for you. Part two teaches you to put on your thinking cap. You do not share your brain with your friends, do you? So be smart. Use your brain to think before you do what friends ask.

When friends try to talk you into doing something with them, they may tell you it will be okay. But you must think if it breaks a law. If it breaks a law, then it is wrong. What are some things that break laws that we should not do? Stealing, taking drugs, or breaking windows with rocks would break laws. You could get arrested for breaking laws. And sometimes you could get hurt, too. If you take drugs, then you could hurt your body. They also can make you act stupid. Drugs are very bad for you.

You must also think if what your friend wants you to do breaks a rule. If it breaks a rule, then it could get you in trouble if you do it. Things that break rules could be lying to your parents about where you are going, cheating on a school paper, fighting, or going someplace where you are not supposed to go. You get in trouble with your parents or teachers if you break rules.

Laws or rules are made to protect us from harm. We should follow them. If we do not follow them, we could get arrested, get hurt, or lose privileges. No one wants those things to happen.

If you break laws or rules, then you will have a sad face. If you obey rules or laws, then you will have a happy face. You will be happier if you stay out of trouble!

A PERSON WHO BREAKS LAWS AND RULES
A PERSON WHO OBEYS LAWS AND RULES

Nicholas says, "Mandy likes to bark. Every day when the mailman comes, she barks at him. She always gets in trouble. We are not supposed to bark in the house. She tries to get me to bark, too. But I know that breaks a rule. And I could lose a privilege if I don't mind Mom. I want to have a happy face. So I do not break laws or rules."

Part 3: Say No to Trouble

The most important part of **Too Smart for Trouble** is knowing what to say or do to trouble.

Here are six different ways to get away from trouble:

1. Just say no
2. Leave
3. Make a joke
4. Suggest a better idea
5. Return the dare
6. Tell an adult that you trust

Let's look at each of these ways to say no.

1. Just Say No

When a friend asks you to ride your bicycle too far from home or to fight someone, you might just say no. You have to sound like you mean it when you say it. But you can still be polite.

There are a lot of ways to say no. Here are a few:

"No, thank you."

"No way!"

"That's wrong."

"Nope."

Shake your head no.

"Are you kidding!?"

"Forget it."

You must look your friend in the eye, stand up tall, and firmly say it. If your friend keeps inviting you to do bad peer pressure, walk away after saying no twice.

2. Leave

If your friends are smoking cigarettes or talking about going in an empty house, you could just leave. Walk on home. Or walk over to another friend's house. Call home for a ride if you are too far from home. It is important to get away from the bad peer pressure.

Don't walk away looking scared or stuck up. You can be friendly. You walk away proud—eyes straight ahead. Even if your friend begs, do not walk back to the trouble. Walk away from the trouble quickly.

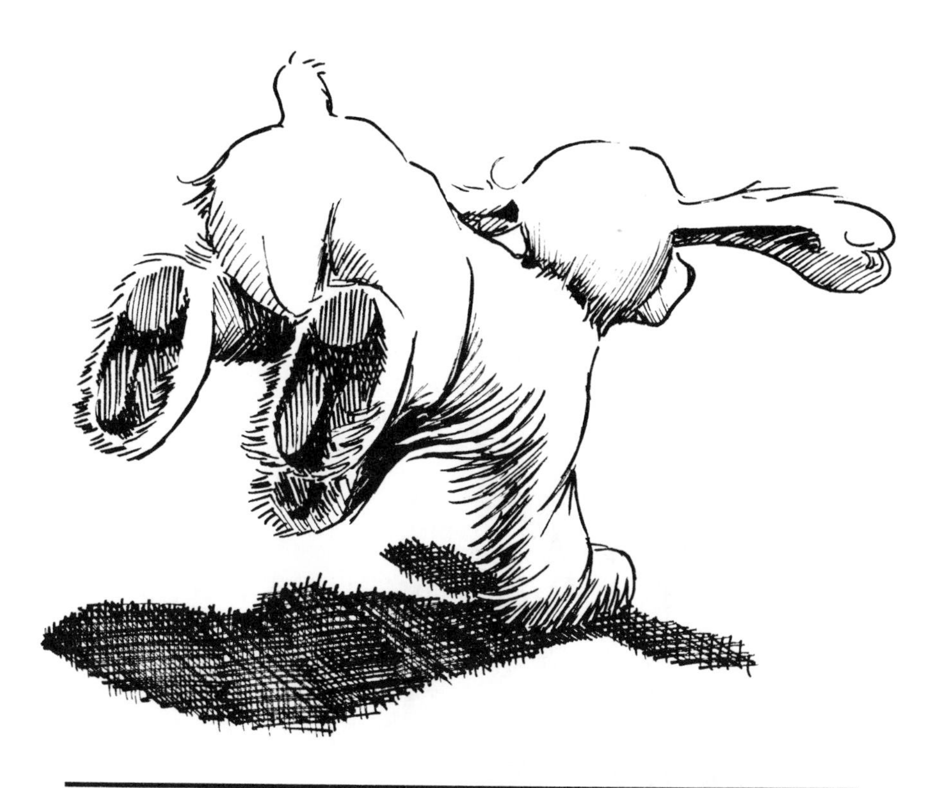

3. Make a Joke

You can also say no to trouble in silly, funny ways. You can make your friends laugh, and you can say no to trouble, too.

If you're asked to go somewhere you shouldn't, say —

"I can't. I've got to go home and straighten up my sock drawer." Or,

"I already have plans. Tonight I've got to brush my dog's teeth." Or,

"I need to spend more time with my plants. Sorry." Or,

"I wish I could, but I've got to walk my goldfish."

When you're asked to do something you shouldn't, say —

"I'm taking flying lessons. I've got to go meet my pilot." Or,

"I've got a new pet. He's gray, weighs two tons, and has a long trunk. It's time to water him again. Bye."

If you are asked to smoke a cigarette, you could say:

"It makes my teeth yellow." Or,

"Ugh. I don't want to stink!" Or,

"No thanks. My lungs asked me to keep them clean."

If you are asked to drink alcohol, you could say:

"I might forget where I parked my brain." Or,

"Not my brand." Or,

"I'm allergic. It makes my skin turn green."

If you are asked to use harmful drugs, like marijuana, you could say:

"I just had some M & M's. That's enough for me." Or,

"I might throw up on you if I take that stuff." Or,

"I'm busy. I have to go watch a commercial." Or,

"I have to meet an astronaut and look at some rocks from the moon."

I bet you can think of other jokes to say no to trouble.

4. Suggest a Better Idea

Another thing you can say to bad peer pressure is to suggest a better idea. Quickly think of something else to do that will not get you in trouble.

Some better ideas are:

> *"I've got a better idea. Let's go play my new game."* Or,
>
> *"Why don't we play some football."* Or,
>
> *"It's time for me to go home. I want to see what's on TV."*

It is important to walk toward the better idea when you suggest it. Then you look serious. Your friend will probably follow you to do the better idea. Smart people can always think of something better to do!

5. Return the Dare

Sometimes someone might dare you to do something wrong. You can return the dare right back.

When someone calls you "chicken," don't say, "No, I'm not." That will not help. That person will just keep daring you. It would be better to say some of these things:

"I'd rather be a chicken than a dead duck." Or,

"If I'm the chicken, you're the egg." Or,

"What's wrong with chickens?" Or,

Just walk away flapping your arms like a chicken.

If someone calls you a "sissy" or "scared," you could say any of these:

"Yes. Anybody smart would be." Or,

"Thanks. I'm glad you noticed." Or,

"You do it. It's your idea. But I think you know better. Bye."

When friends say they will not be friends with you if you don't do something wrong with them, you can say any of these:

> *"If you were my friend, then you would not try to boss me."* Or,
>
> *"Yes, I am your friend. And that is why I'm not going to do this with you."*

There are lots of ways to know what to say to dares.

6. Tell an Adult

And when you're asked to do something wrong, you can always tell an adult. There might be times when a friend suggests something really bad. Someone might get hurt. You should quickly tell your parents or their parents. If trouble happens at school and you don't know what to do, ask your teacher.

Sometimes a friend may ask you to keep a secret. There are good and bad secrets. A good secret is fun, like a surprise party. A bad secret could cause someone to get hurt, feel bad, or get in trouble. Always tell bad secrets to an adult that you trust. The adult will help you decide what to do.

Nicholas says, "Wow! You have now learned six things to say to bad peer pressure: say no, leave, make a joke, suggest a better idea, return the dare, and tell an adult.

"No matter which way you say no, do it quickly. Try to get away from bad peer pressure in 30 seconds. Thirty seconds is fast, too. That is less time than it takes for you to watch a commercial. If you take too long, you might get in a fight. Or you might give in. So get away fast!"

Chapter 5

Let's Practice!

Debra's Bad Peer Pressure

Debra was at a sleepover at Stephanie's house. Stephanie was nine today. They had a fun party. It was now bedtime. Stephanie's mom told the girls to go to sleep.

The girls were in bed. But they were too excited to go to sleep. They kept giggling.

Stephanie whispered, "Debra, let's wait until my mom goes to sleep. Then we'll get up and play some more. She won't find out."

If you were Debra, what would you say or do?

Debra's Good Decision

Debra looked and listened. She noticed that Stephanie was acting sneaky. She knew it was bad, because they would be breaking a rule and lying, too.

So Debra suggested a ***better idea***. She said, "Let's go to sleep. Then we can get up early and play some more." Debra did not want to get in trouble. She knew her mother would not let her go to a sleepover again. She followed the rules. Debra is **Too Smart for Trouble!**

George's Bad Peer Pressure

Derrick was visiting his friend, George. Derrick's mom was taking a nap. The boys were bored. They could not think of anything fun to do.

Finally, Derrick said, "My dad has a gun in the drawer. Let's look at it."

If you were George, what would you say or do?

George's Good Decision

George thought to himself, "This is bad peer pressure. Someone could get hurt or killed."

So George said, "No, we could get hurt. I have a better idea. Let's see what's on TV." He suggested a ***better idea*** and started walking toward the TV. He decided to ***leave*** the room.

If Derrick does not follow him, then he should go tell Derrick's mom. George decided this was serious. He ***got help from an adult.*** George is **Too Smart for Trouble!**

Kim's Bad Peer Pressure

Kim and Sandy were in the first grade. They lived on the same street. They always walked home from school together.

One day on their way home, a car pulled up next to them. The man asked them a question. They did not know him.

Sandy started to walk toward his car to see what he wanted. If you were Kim, what would you do?

Kim's Good Decision

Kim knew not to talk to someone that she did not know. It does not matter if the person wants to give you candy, or show you a puppy, or ask you a question. Kim knew ***never*** to talk to strangers. It does not matter if the stranger is a man or a woman. If you do not know the person, do not talk to him or her. Do not get close to the car. The person might be mean. It could be dangerous!

So Kim quickly reached for her friend's hand and said, "Sandy, no! We don't know him! Let's hurry home!" Both girls ran quickly toward home. They ***left*** fast! And they ***told their parents*** what had happened when they got home. Kim is **Too Smart for Trouble!**

Juan's Bad Peer Pressure

Juan and David were at recess. They had been playing kickball. When the game ended, David asked Juan to go around the corner of the building.

Juan asked, "Why?"

David said, "To smoke some cigarettes, man. It's cool."

If you were Juan, what would you say or do?

Juan's Good Decision

Juan told his friend no with a ***joke.*** He said, "Cigarettes taste terrible. I might 'barf' on you if I smoked one!"

Juan said no in a funny way. David will still be his friend. And David knows he can't boss his friend around. Juan is **Too Smart for Trouble!**

Maggie's Bad Peer Pressure

Maggie got to school early one day. She sat at her desk drawing. One of her friends, Shannon, came over.

Shannon said, "Do you like Linda?"

Maggie said, "Yes, I think she's nice."

Shannon said, "Well, I don't like her. She's fat, and she wears ugly clothes."

Maggie said, "Shannon, maybe those are the best clothes she has. Don't make fun of her."

Shannon then said, "Maggie, if you want to be my best friend, then you won't talk to Linda any more."

If you were Maggie, what would you say or do?

Maggie's Good Decision

Maggie decided to ***return the dare.*** She said, "Shannon, if you were my friend, you wouldn't try to tell me whom to talk to. I like you a lot, and I want to keep you as my friend. But Linda needs people to be nice to her. We don't have to be mean."

Maggie knew that she could have more than one friend. She does not want friends to decide for her whom she should talk to. That's her decision.

Sometimes people are not kind to others because they don't feel good about themselves.

Finally, Maggie said, "Shannon, let's talk about something else. What do you want to do this weekend?" Maggie came up with a ***better idea.*** The two girls had fun making their plans. Maggie is **Too Smart for Trouble!**

Jeremy's Bad Peer Pressure

Ronnie and Jeremy were in the restroom at school. Ronnie said, "I've got a great idea. Let's stop up the commode with toilet tissue. We won't get caught."

If you were Jeremy, what would you say or do?

Jeremy's Good Decision

Jeremy thought quickly. He knew this broke a school rule and the law. He did not want to get in trouble. He also thought this was a really stupid idea!

So Jeremy just ***left.*** He walked back to class and started doing his work.

Jeremy is **Too Smart for Trouble!**

Molly and Karen's Bad Peer Pressure

It is Saturday, and Molly has gone next door to visit her friend, Karen. Karen has an older sister, Nicole, who is in eighth grade. Nicole was babysitting them.

Molly and Karen had finished playing a game. They walked into Nicole's room. Nicole and another girl were smoking a funny smelling cigarette.

Karen did not know her sister smoked. She asked Nicole what she was doing.

Nicole said, "Smoking pot, grass, marijuana. It's all the same stuff. It's good. Wanna try some?"

If you were Molly and Karen, what would you say or do?

Molly and Karen's Good Decision

Molly and Karen said ***no.*** Then they ***left*** and went back to Karen's room. They talked about this for a long time. They decided to ***tell an adult.*** Both girls told their parents. This was serious. They knew that Nicole could get sick or hurt using drugs. And she could be arrested. They liked Nicole. So they told their parents. Now Nicole's mom knows about the drugs and can help her quit using them.

Molly and Karen are **Too Smart for Trouble!**

Robert's Bad Peer Pressure

Scott and Robert were talking about the new boy down the street.

Scott said, "That new kid acts stuck up. I don't like him. Let's push him off his bicycle when he rides by."

Robert said, "No."

Then Scott said, "Are you chicken?"

If you were Robert, what would you say or do?

Robert's Good Decision

Robert knew that Scott was trying to dare him to be unkind. He knew that was wrong. He thought the new kid might even be a nice guy.

So he said, "Scott, if I'm a chicken, then you're the egg!" He ***returned the dare.***

Scott said, "Fine. Just forget it." He acted mad at Robert. But he only stayed mad a few hours. Robert knew that a real friend does not stay mad long.

Robert is a kind person. And he is **Too Smart for Trouble!**

Tonya's Bad Peer Pressure

Shirley lived in an apartment. Her friend, Tonya, was visiting. It was hot. The girls got permission to go outside and play.

Shirley said, "I know how we can cool off. Let's go to the apartment swimming pool. We can sit on the side and put our feet in the water."

If you were Tonya, what would you say or do?

Tonya's Good Decision

Tonya said, "Are you kidding? What if we fell in the water? We could drown. No way."

Tonya was serious when she said ***no.*** It made Shirley think.

Shirley said, "You're right. It was a dumb idea."

Tonya knew not to go near the swimming pool without an adult. Tonya helped herself. And she helped a friend. Tonya is **Too Smart for Trouble!**

Jeff's Bad Peer Pressure

Doug and Jeff rode their bikes to the park at the end of the street. They were playing catch. There were some older boys in the park. The older boys were sitting and talking.

Two of the older boys came and said, "Hey, you want some beer?"

Both Doug and Jeff said, "No, thanks."

The boys started teasing Doug and Jeff. Doug finally said, "I'll try just one sip."

If you were Jeff, what would you say or do?

Jeff's Good Decision

Jeff had to think quick. He did not want to get in a fight with the older boys. He decided to make a ***joke.*** He said, "Sorry, that's not my brand of beer." The older boys laughed. Next he had a ***better idea.*** He said, "It's time for me to get home. Come on, Doug." Doug followed Jeff home. They ***left.***

Wow! Jeff used three ways of getting out of trouble. Jeff is **Too Smart for Trouble!**

Nicholas is proud that he, Shawn, Mandy, and Cedric have all now learned to be **Too Smart for Trouble**. They are all making good decisions and they have lots of friends, too. They will always remember to think for themselves.

I am sure you now know more what to say or do in each story you just read. I am so excited about how much you have learned! And I know you are proud of your good decisions. If you keep making good decisions, then you can stay out of trouble. And you may be able to even help others.

Nicholas and I want you to know that we think you are very special. You are **Too Smart for Trouble!**

Letters From Paw Pals

Benji

March 7, 1990

Dear Nicholas:

Three barks for your efforts to keep kids out of trouble!

I've spent my career on stage and silver screen trying to bring hope and joy to young and old alike, and it's always saddened me to hear about kids in trouble because of negative peer pressure, and to meet their broken-hearted parents.

Your book teaches responsible friendship and I know it will bring help and joy to many of our young friends.

Keep up the good work, and God bless.

Woof,

7720 Glen Albens · Dallas, Texas 75225 · 214-368-2551

Editor's Note: Benji™, America's Most Huggable Hero, soared to worldwide superstardom in 1975 with the release of his first motion picture. Since then, he has starred in three more top-grossing motion pictures, four network television specials and a network series. He is currently at work on another film.

August 20, 1989

Nicholas
c/o Ms. Sharon Scott

Dear Nicholas,

I am a dog named "Wally" and I work for the Dallas Police Department. I help the Dallas Police find illegal drugs just by using my nose!

I just finished pawing through the book you helped Sharon Scott write, TOO SMART FOR TROUBLE. I hope the children, parents, and teachers who read this book like it as much as I did. Sometimes it's hard for kids to do the right thing when their friends are doing the wrong thing. Your book will keep kids from barking up the wrong tree. Keep up the good work.

Sincerely,

MACK M. VINES
CHIEF OF POLICE

POLICE DEPARTMENT POLICE AND COURTS BUILDING DALLAS, TEXAS 75201

S/N 753-027-318
POL-00152

Editor's Note: Wally is a yellow Labrador Retriever. He and his partner, Senior Corporal James F. Hughes, have been involved in 50 drug arrests and have seized $100 million in illegal drugs. They are assigned to the Drug Enforcement Administration Task Force at Dallas/Fort Worth International Airport.

RONNIE F. CLACKUM
Director

Police Headquarters Building
7930 North McDonough Street
Jonesboro, Georgia 30236
(404) 477-3600

August 4, 1989

Nicholas
c/o Ms. Sharon Scott
Garland, Texas 75043

Dear Nicholas,

We are Clayton County Police Drug Detection Dogs, "Buster" and "Magic". We have never met you, but we just had to write and tell you how excited we are about the new book you co-authored with Ms. Scott. TOO SMART FOR TROUBLE will be a big help to us when we teach children to avoid trouble and be safe. NO is an easy word to say, but sometimes it is hard to know exactly HOW to say it.

When we are not busy helping our partners find illegal drugs, we teach children to avoid trouble and be safe. We help thousands of children learn to say NO, but every year a few of them still get into trouble because they don't know HOW! Your book will really help them and give parents, teachers and other adults a way to teach children HOW to say NO!

We have to get back to work now - but before we do we want to thank you for TOO SMART FOR TROUBLE. Please visit us the next time you are in Georgia.

Your Friends,

Buster and Magic

B&M:ej

Editor's Note: Buster and Magic are black Labrador Retrievers. They, along with their partners, Sergeants Dennis Messman and Toni Tidwell, have been involved in the arrests of over 150 drug dealers. Their enforcement activity has resulted in the seizure of $10 million in illegal drugs.

Ending Note to Parent/Teacher:

Let ***Too Smart for Trouble*** be just the beginning of lots of practice with your child or student. Suggested ways to practice:

1. Each of the three parts of ***Too Smart for Trouble*** can be practiced. Example, ***"Look and Listen"*** could be practiced by asking the child to look for any clues to trouble for that day and report them to you later. This makes the child practice observing and listening skills and open discussion. Part two, ***"Think: Is It Good or Bad?"*** could be practiced by giving hypothetical situations and asking what good or bad might happen if the child did them. This allows the child to practice evaluating and deciding. Part three, ***"Six Ways of Saying No"*** could be recited. You could also have the child draw pictures about these six ways of avoiding trouble.

2. Educate yourself on the harmful effects of the "gateway" drugs: tobacco, alcohol, and marijuana. It is important that you have some simple, factual, and current information in each of these areas so that you can begin early education on their harmful effects. It must begin early since these decisions are occurring at younger ages each year (i.e., the average age for youth's decision about first alcohol usage is now 11 1/2 years).

3. Evaluate your own habits and your own ability to "say no" to adult peer pressure. The "do as I say, not as I do" philosophy just does not work. Young children often ask me how to get their parent to quit smoking. They are fearful that they will die and leave them. Children have enough fears without

our adding more to them. Not only that, but they also imitate our behaviors.

4. Write the following organizations concerning children's magazines or adult newsletters. All have inexpensive membership dues or costs. Each provides excellent information:

   National Parents' Resource Institute for Drug Education, Inc. (PRIDE)
   The Hurt Building, Suite 210
   50 Hurt Plaza
   Atlanta, GA 30303
   1-800-241-7946

   National Federation of Parents for Drug-Free Youth
   1423 N. Jefferson
   Springfield, MO 65802-1988

   "Winners" Magazine
   C/O Narcotics Education, Inc.
   12501 Old Columbia Pike
   Silver Spring, MD 20904

   "PTA Today" Magazine
   C/O The National PTA
   700 North Rush St.
   Chicago, IL 60611-2571

   Committees of Correspondence
   57 Conant St., Room 113
   Danvers, MA 01923

PRIDE Canada, Inc.
Suite 111, College of Pharmacy
University of Saskatchewan
Saskatoon, Saskatchewan S7N0W0
CANADA
1-800-667-3747

5. Most importantly, do lots more practice of the ***Too Smart for Trouble*** skills. Set up role-play skits to give the child the opportunity for more practice before the "real thing." Tell the child the location, then pretend you are the pressuring peer. A skit might look like this:

Parent: "Johnny, let's practice a **Too Smart for Trouble** skit. I'll pretend I'm a friend of yours, and we are outside riding our bicycles. Think quickly and see if you can decide what to say or do."

Johnny: "Okay, this will be fun."

Parent: "The neighbors are out of town. Why don't we go in their garage? We can look around. Maybe there is something to play with."

Johnny: "No, that's not a good idea."

Parent: "Yes, it is. It'll be fun. They'll never know."

Johnny: "Forget it! I'm thirsty. I'm going to get a drink." (Johnny walks away.) "You want to come?"

Parent: "Yeah, I guess."

Parent: "Johnny, that was fantastic! Which of the ways of saying no did you use?"

Johnny: "I said no and came up with a better idea."

Parent: "And you walked away also. I am so proud of how quickly you thought. You are **Too Smart for Trouble!"**

These skits can be practiced:

1. In the car to or from school
2. At bath time
3. By having your child teach, with your help, a younger brother or sister
4. With neighborhood practice groups of children and parents
5. At meal time
6. While on vacation
7. To show off for a visiting relative
8. At scout troop meetings
9. As a bedtime story
10. Really anywhere!

The point, of course, is the more practice your children have, the less likely they will be to yield to negative peer pressure.

If your child relies on *only* one of the six ways of saying no, you might ask the child what other ways there

are. This helps the child build a variety of effective responses.

As your child gets older, you may want to have him or her read the preteen/teen version of this book, ***How to Say No and Keep Your Friends.*** It presents more ways of saying no in the more difficult situations that older youth face.

For information on ordering a ***Too Smart for Trouble*** poster, featuring Nicholas, write Sharon Scott and Associates.

Sharon Scott is available in the following capacities:

* Keynote presentations at conferences
* Workshops for youth and parents on ***Too Smart for Trouble*** and other skills
* Developing positive peer groups
* Inservice for teachers and counselors
* "Positive Parenting" columns for newsletters
* Video for teens

Write for further information:

Sharon Scott (and Nicholas)
Windy Star Ranch
P.O. Box 6
Weston, TX 75097-0006

For information on ordering Sharon Scott's books:

* ***Peer Pressure Reversal: An Adult Guide to Developing a Responsible Child***
* ***How to Say No and Keep Your Friends***
* ***Positive Peer Groups***
* ***When to Say Yes! and Make More Friends***
* ***Too Smart for Trouble***

Contact:

Human Resource Development Press
22 Amherst Rd.
Amherst, MA 01002
1-800-822-2801 (outside Massachusetts)
(413) 253-3488 (within Massachusetts)

Discounts given on quantity orders.

CANTON PUBLIC LIBRARY
3 1631 00185 4459

Sanborn, and Leah Wohlin. Our interviewees were unfailingly generous. All are cited and have reviewed and approved or amended what is attributed to them. I especially thank Viraf Soroushian for actively participating from afar in my research about UCSD, and Claire Tindula for putting me in touch with all the right people.

TransFair USA (now Fair Trade USA) graciously hosted me as a volunteer/visiting scholar in August 2009. Arnab Basu, Shareen Hertel, and Karla Utting generously contributed figures and unpublished information about their research. For invaluable financial support I thank the Chris and Warren Hellman Foundation and UCSD's Academic Senate. Melanie DuPuis, Jeff Haydu, Katherine Mooney, two anonymous reviewers selected by the University of Washington Press, and particularly my sponsoring editor, Lorri Hagman, deserve special appreciation for skillfully guiding my many revisions.

Writing a book can be a lonely process. It has been immeasurably helpful to connect with academic, activist, and personal supporters. These include many of the individuals mentioned above, but also others who were not involved in the manuscript's evolution but have nevertheless contributed to its fruition. Many thanks to Lance Bennett, Mary Blair-Loy, Miguel Centeno, Maria Charles, Abigail Cooke, Sara Curran, Edgar Kiser, the Linton family, Isaac Martin, and especially Gary Morgan, who patiently listened to way too many "book stories" and never complained.

FAIR TRADE FROM THE GROUND UP

Introduction

On a busy Saturday in the summer of 1993, a group of protesters organized by the US/Guatemala Labor Education Project (US/GLEP) picketed in front of Starbucks' recently opened 200th store in Seattle's historic Pioneer Square. They wanted Starbucks to stop purchasing coffee from plantations where workers were treated poorly and not paid fair wages. The company's coffee buyer appeared on the scene and conversed earnestly with some of the protestors. Later that day he had tears in his eyes when he told me—the store's manager—that the demonstrators were right: the Central American businessmen that the buyer dealt with had little regard for the workers who tended their plantations or for the smallholder farmers to whom they paid a pittance for unprocessed coffee cherries. But why had activists targeted Starbucks instead of the multinational corporations that control the vast majority of the coffee industry? It would be impossible for Starbucks to meet US/GLEP's demand because the small company (remember, this was 1993) almost always purchased beans from brokers rather than directly from producers. There was no way to pinpoint where the coffee came from.

Fast-forward to 2011. The former coffee buyer has been one of Starbucks' senior vice presidents for quite some time, in various capacities, including vice president for social responsibility. As for me, after leaving the company in 1996 to pursue my graduate education at the University of Washington, I thought about coffee mainly as a pleasurable caffeine source, although I was aware of "Fair Trade" coffee's launch in the United States,[1] that the activist group Global Exchange had—by threatening to demonstrate at stores across the country—successfully lobbied Starbucks to buy some of it, and that the company had established some preliminary sourcing guidelines in an attempt to address mounting concerns about workers and the environment. Then, in 2001, the founder of the US Fair Trade labeler TransFair USA (now Fair Trade USA) visited the University of Washington campus, accompanied by a Central American coffee farmer. This was my real introduction to Fair Trade as a way for smallholder farmers to organize democratically, capture more of the export value of their produce, and sell it in a market that guarantees a floor price. I was surprised to learn that, at the time, Fair Trade activists had done a much better job of organizing and certifying coffee farmer cooperatives than of creating a market for the coffee they produced. Less than half of it was being sold under Fair Trade terms. Under the mentorship of political scientist Margaret Levi, a posse of undergraduate honor students and I set out to learn why such a large gap existed, and what could be done to narrow or close it. Our research yielded suggestions about how to better integrate Fair Trade into the specialty coffee sector, chronicled winning strategies employed by student- and faith-based campaigns, and proposed ways to institutionalize Fair Trade coffee (i.e., make it the only option) in large public and corporate facilities (Levi and Linton 2003; Linton, Liou, and Shaw 2004). It set me on the path that has led to this book.

Pragmatic and sociological questions motivate this work. Is the Fair Trade movement living up to the hopes of its early and current advocates? What strategies seem most likely to succeed in further expanding Fair Trade's market share? How can a solidarity and social justice–based movement best negotiate growth that involved "mainstreaming" Fair Trade to include businesses such as Wal-Mart? Sociologists often distinguish Fair Trade as a transnational social movement that seeks to mobilize people in the global North on behalf of other people they will probably never meet. As such it is quite promising because it offers a ready (albeit not all-encompassing) solution to the problem of unequal relationships within networks of global trade (Keck and Sikkink 1998). But how does Fair Trade construct solidarity between farmers, activists, businesses, and consumers—or fail to do so?

Fair Trade is one of many connected "new" social movements based on NGOs (non-governmental organizations) and networks as opposed to trade unions or political parties. Along with organic farming, living wages, the preservation of indigenous cultures, animal rights, localized food sovereignty, and the protection of biodiversity, Fair Trade is part of a larger push for sustainability and against the exploitation of people, animals, and the environment. These movements are "new" because their success or failure largely depends upon conscious consumption rather than organized boycotts or direct political action (Micheletti 2003). How does the Fair Trade movement exert leverage when conventional, state-centered strategies are not feasible? I do not have all the answers, but I hope to inform ongoing evaluations and discussions. My research, and my reading of others' research, strongly suggests that Fair Trade is changing norms for producers, businesses, and consumers. But the size and scope of its ultimate impact depends upon the successful navigation of ongoing tensions between movements and markets. For example, does it hurt the movement when Fair Trade certified goods are sold in mainstream businesses that otherwise do not uphold fair labor standards? Is this a sellout or an important way to generate considerable new revenues for farmers and to expose a new set of consumers to Fair Trade products? I will argue that the social-movement and market-centered parts of Fair Trade are complimentary to the degree that there is transparency for all stakeholders.

A BRIEF HISTORY

Fair Trade is rooted in three beliefs:

- We are all connected. Consumers have the power to express solidarity with producers.
- Existing world trade practices affect underdevelopment and the unjust distribution of wealth among nations.
- Paying producers in less-developed countries a fair price for their products is a more efficient way of encouraging sustainable development than aid.

In April 1985 Paul Katzeff, the owner of Thanksgiving Coffee Company in Fort Bragg, California, accepted an invitation to visit La Unión Nacional de Agricultores y Ganaderos (UNAG), Nicaragua's pro-Sandinista coffee organization. "I hung out with the Sandinistas in the mountains, where they were fighting the Contras. I met with three *commandantes* of the Revolution. I was educated about the relationship between coffee and revolution" (quoted in Pendergrast 1999, 354). Returning to California, Katzeff changed his company's slogan to "Not Just a Cup, but a Just Cup" and labeled the Nicaraguan beans he roasted as "Coffee for Peace," donating fifty cents per pound to the Sandinistas. When the Reagan administration banned the import of all Nicaraguan goods, Katzeff got around the embargo by having his Nicaraguan beans shipped to Canada and roasted there so that it could be traded as a product of Canada.

A year later, Rink Dickinson, Jonathan Rosenthal, and Michael Rozyne—who had met while working in a New England food co-op—formed the organization Equal Exchange. "We aim to create a process," wrote Rosenthal, "that allows people to reconnect with the people who grow much of the food and with the ecology from which it comes" (quoted in Pendergrast 1999, 355). Their first product was "Café Nica," Nicaraguan coffee brought to the United States via the Netherlands (Equal Exchange 2008). Companies and Alternative Trade Organizations (ATOs) such as Thanksgiving Coffee and Equal Exchange built a small niche for socially conscious coffee in the United States.

Fair Trade's deep roots in political solidarity and cooperativism are still apparent even though the movement has evolved into a market-based approach to promoting social, economic, and environmental sustainability (see, e.g., Le Velly 2006, Low and Davenport 2005, VanderHoff Boersma 2009). The contemporary movement originated in 1988, grounded in the premise that prices set by exploitative trading systems were unjust and that an alternative value chain that incorporated fairness and sustainability was in order (Raynolds 2002). The first Fair Trade labeling initiative was Max Havelaar Netherlands, established via collaboration between the Mexican cooperative Unión de Comunidades Indígenas de la Región del Istmo (UCIRI) and Dutch ethical trade activists, orchestrated and mediated, for the most part, by two people: farmer/priest/academic Franz (Francisco) VanderHoff Boersma and economist Nico Roozen. They launched the Max Havelaar Fair Trade seal, which was offered to mainstream coffee companies if they agreed to buy some—even a small fraction—of their coffee on Fair Trade terms. By paying a certification fee, roasters purchased the right to apply the seal to an amount of their roasted coffee equivalent to the amount of Fair Trade coffee they purchased (Talbot 2004, 203-4). Max Havelaar's strategy of targeting larger, profit-driven roasters and getting Fair Trade coffee into supermarkets greatly expanded consumers' exposure and access to ethically traded coffee.

With Max Havelaar as a model, other European labeling initiatives soon followed. Some of these, including the Fairtrade Foundation in the United Kingdom, were founded with the help of antipoverty organizations such as Oxfam, World Development Movement, and Christian Aid. In 1998 Paul Rice launched the US Fair Trade labeler TransFair USA. Rice had worked as a development specialist in rural Nicaragua and founded Promotora de Desarrollo Cooperativo de las Segovias (PRODECOOP), a Fair Trade and organic cooperative representing almost 3,000 smallholder coffee farmers in northern Nicaragua. PRODECOOP's success in capturing European Fair Trade and gourmet coffee markets introduced Rice to "the power of Fair Trade as an innovative strategy for grassroots empowerment and sustainability" (Fair Trade USA 2011a). Establishing an NGO to open the US specialty coffee market to Fair Trade farmers was a logical next step.

All of the country-specific Fair Trade labelers are NGOs that audit transactions between importers and the Fair Trade certified producers from whom they source. Their efforts to directly link producers and potential buyers are crucial to establishing and expanding Fair Trade markets because it is via these relationships that farmers develop practices that yield the qualities (taste, appearance, size, and so on) that importers are interested in. The same NGOs provide the labels that differentiate Fair Trade products, and promote these products by educating consumers and working with retailers, institutional purchasers, and local activists.

The Fair Trade movement first focused on coffee because coffee is an important source of foreign exchange for most producing countries and because almost half of the world's coffee is grown by smallholder farmers. In the first decade of the US movement, a global "coffee crisis," rooted at the end of the International Coffee Agreement of 1989, was fomenting. Temporary price spikes due to crop damage in Brazil encouraged over-planting of new coffee trees as well as new entrants into the market, notably Vietnam. Coffee production worldwide was increasing faster than its consumption (see, e.g., Bates 1997, Talbot 2004). In 2002 Oxfam reported:

> There is a crisis affecting 25 million coffee producers around the world. The price of coffee has fallen to a 20-year low and long-term prospects are grim. Developing-country coffee farmers, the majority of whom are poor smallholders, now sell their coffee beans for much less than they cost to produce. The coffee crisis is becoming a development disaster whose impact will be felt for a long time. (Gressler and Tickell 2002, 6)

Producers who were integrated into Fair Trade networks were able to weather the crisis, and since then Fair Trade has seen phenomenal growth. Worldwide sales increased 40 percent between 1999 and 2004. Between 2004 and 2007, the average *yearly* increase was 47 percent. Sales went up by another 22 percent between 2007 and 2008, and 15 percent between 2008 and 2009 (FLO 2004, 2008, 2010a). This trajectory has added billions of dollars to the incomes of farmers in the developing world, benefitting more people each year as existing Fair Trade markets grow and new products enter the mix. The rapid growth of Fair Trade has also generated controversies over standards, monitoring, and retailing, which several sections of this book address (see, e.g., Fridell 2007; Jaffe 2007; Raynolds 2007; Renard 2003, 2005). What is fair? Who should shoulder the cost of certification and monitoring? How committed to Fair Trade must a retailer be in order to use the label?

FAIR TRADE TODAY

> Fair Trade is a trading partnership, based on dialogue, transparency and respect, which seeks greater equity in international trade. It contributes to sustainable development by offering better trading conditions to, and securing the rights of, marginalized producers and workers—especially in the South. Fair Trade organizations (backed by consumers) are engaged actively in supporting producers, awareness raising and in campaigning for changes in the rules and practice of conventional international trade. (Definition crafted in 2001 by FINE, an umbrella organization of Fair Trade networks; cited in Mohan 2010, 19-20.)

Since 1997, country-specific Fair Trade NGOs have been united under the umbrella of Fairtrade International (formerly the Fairtrade Labelling Organization and still abbreviated as FLO). Fairtrade International sets Fair

Trade standards and coordinates the activities of country-specific Fair Trade organizations. FLO-CERT GmbH is an independent certifying agency that performs inspections to ensure that producers, traders, and retailers comply with Fair Trade's criteria.[2]

The current generic standards for Fair Trade certification are:[3]

* Smallholders organize in democratically governed cooperatives (so they can sell directly to importers). These may be second-tier organizations that encompass several smaller cooperatives.
* Plantations pay at least legal minimum wages; provide good housing where relevant; meet established health, safety, and environmental standards; and guarantee their workers' right to join unions. No child labor or forced labor may occur.
* *Producers and certifiers* collaborate to establish progress requirements. These encourage producers to continuously improve working conditions and product quality, and to increase the environmental sustainability of their activities.
* In addition to paying a set price to producers that is intended to cover the costs of sustainable production and living, traders pay a fixed social premium that producers collectively decide how to allocate.
* Traders must extend credit to producers if they request it, and are encouraged to enter into contractual arrangements, thus allowing for long-term planning and sustainable production practices.[4]

To summarize, Fair Trade is a market-based approach to integrating social responsibility and global commerce. It offers a stable price and allows small farmer cooperatives and sometimes plantations in the global South to directly export their coffee, tea, cocoa, fruit, and other produce rather than participate in conventional value chains with many intermediaries. Thus producers are able to capture significantly more of their crops' export value. Fair Trade buyers also pay a "social premium" to producer groups; these funds are apportioned as members see fit.

Some producers also realize indirect benefits from Fair Trade. For example, in their study of Fair Trade coffee cooperatives in Nicaragua and Tanzania, sociologist Gautier Pirotte and his colleagues point out that the cooperatives nurture a sense of solidarity just as they train producers and advocate an entrepreneurial spirit. "Fair Trade has given small-scale producers the economic security to enable them to develop and take charge of their own lives within the cooperative network" (Pirotte, Pleyers, and Poncelet 2006, 450). In a similar vein, the marketing director of a South African Fair Trade winery told me that, by empowering workers, Fair Trade was bringing them to think of farming as a profitable business with a future instead of an impoverished lifestyle.[5] Although still a relatively small phenomenon, Fair Trade's continued growth illustrates its potential as a strategy for making trade liberalization feasible for farmers in developing countries.

ABOUT THIS BOOK

The Fair Trade movement has sparked the interest of social scientists and development professionals, in part because scholars of social movement and labor debate the extent to which transnational campaigns can truly empower smallholder farmers and workers in the global South while appealing to corporations and consumers in the global North (Anner and Evans 2004; Bartley 2003; Fung, O'Rourke, and Sabel 2001; Jaffee 2007; Keck and Sikkink 1998; Seidman 2007; Shreck 2005). We now have a substantial body of research that looks specifically at aspects of Fair Trade: field studies, marketing or consumer research, and critiques.[6] This book aims to provide an accessible synthesis of the existing literature as well as to contribute to it. It documents and evaluates Fair Trade's achievements to date, in terms of promoting sustainable development and human rights in the global South and new norms of consumption in the global North. My main objective is to increase our understandings of how to strengthen Fair Trade and how to negotiate the challenge of creating an alternative to mainstream markets while working within them. To this end, the original research I present pays attention to how producers view their involvement in Fair Trade and to how they are using the Fair Trade social premium. On the consumer side, the book profiles a diverse group of Fair Trade activists in the United States and documents a large school's successful effort to become a Fair Trade university.

The book follows Fair Trade products from their origins to their destinations along what social scientists call global value chains (see, e.g., Daviron and Ponte 2005; Gereffi, Humphrey, and Sturgeon 2006). In a value chain, products accrue monetary and/or symbolic value at each junction. Geographer Abigail Cooke and her colleagues (2008) have suggested extending the value chain perspective to *webs* of value chains—that is, incorporating nodes at which chains intersect. This is important here because most Fair Trade producers are also involved in other, "conventional" value chains and may be selling some of their crops locally (as discussed in chapter 2). Fair Trade does not necessarily replace other value chains; it may simply enhance producers' options.

Fair Trade's broader goal is to change norms of trade and consumption, yet its success is usually measured according to what political scientist Gavin Fridell calls the "shaped advantage" perspective of Fair Trade, in which the main goal is to mitigate the negative effects of unstable markets (2007, 85).[7] Less elegantly stated, Fair Trade's gains or losses are measured in terms of sales and/or market share. I have already presented a snippet of such data, and will offer more. But I am more interested in this question: When, where, and how have Fair Trade's participants maintained a transformative agenda while successfully marketing various products? I argue that Fair Trade is successfully promoting norm change in multiple dimensions, although there are still many challenges and opportunities. Throughout the book I suggest ways to impel the movement in a positive direction.

WHAT IS TO COME

The first part of this book (chapters 1 through 3) considers Fair Trade from the standpoint of the producers; the second part focuses on the consumer side of the value chain (chapters 4 through 7). These two views intersect significantly because producers need to understand their markets, and conscious consumers (including businesses and government institutions) want to know what their money is supporting.

Producers and Products

As Fair Trade's reach expands, its focus remains on crops that thrive in tropical to Mediterranean climates: tea, cacao beans, olives, wine grapes, bananas, tropical and citrus fruits, and spices such as vanilla and cinnamon. These are things that many of us want but cannot grow or buy from local farmers, and they are often produced under ambiguous conditions. Fair Trade aims to make these crops socially and environmentally sustainable by establishing links between faraway farmers and the people who buy their produce.

Chapter 1 reviews the current literature about Fair Trade producers and products. It synthesizes many field studies and value-chain analyses. In its conclusion, I advocate a uniform framework for studies that seek to evaluate Fair Trade's success. The idea is not that every study should follow all aspects of one template but rather that, by incorporating parts of the template, studies conducted in disparate parts of the world and related to various products will become more comparable.

Chapter 2 zooms in on Fair Trade coffee producers in Guatemala. In it, I employ original survey data about leaders and members of Fair Trade cooperatives to explore their understandings of Fair Trade, their views about it, how individual farmers have benefitted from it (or not), and issues that can undermine it (e.g., "side-selling").

Chapter 3 (written with Marie Murphy) zooms out to systematically explore how Fair Trade producers worldwide are using the social premium they receive. We employ a data set constructed from approximately 250 Fair Trade producer profiles, representing most producer groups who were actively selling under fair terms in 2008. The analysis demonstrates that the majority of producer groups are spending at least some of the premium on projects that benefit entire communities, not just themselves.

Consumers

Fair Trade aims to extend equitable and sustainable relationships across international borders, but how does this work in the global North? Like the previous section, this one begins with a comprehensive survey of existing literature. Chapter 4 describes Fair Trade activists' efforts to get businesses to work with them and summarizes several studies about Fair Trade and businesses—from "mission-driven" to "market-driven" firms. It then highlights important research about Fair Trade and consumers that addresses questions such as what makes people express their values when they shop? Is Fair Trade consumption tied to income? In reviewing these studies, my aim is to offer concise, readable summaries without compromising accuracy or muffling the researchers' goals and findings. The chapter concludes with a broader discussion of political consumerism.

Chapter 5 (written with Rebecca Kahn) is based on in-depth interviews with people who are Fair Trade activists within their businesses, workplaces, or communities. This is a heterogeneous group, but it becomes clear that each activist is linking Fair Trade with other efforts to promote sustainability, and that each understands sustainability in social, economic, and environmental terms. Studies about Fair Trade activism have focused almost exclusively on European actors; this chapter provides a US context. It also spotlights Fair Trade Towns, a movement that is well established in the United Kingdom and now emergent in the United States.

Chapter 6 is a case study of a successful campaign to make a large research university a "Fair Trade university." It is informed by current and former students, relevant administrators, and the chancellor. The chapter describes how the campaign was born and its bumpy progress; ultimately it offers a blueprint that other schools could follow. Once again, there is a clear emphasis on connecting Fair Trade to broader-based sustainability efforts.

Fair Trade certification is growing quickly, exerting a positive effect on the lives of more and more producer families and communities. At the same time, Fair Trade is giving consumers an opportunity to exercise moral choice in their purchasing practices and is serving as "an effective critique of business practices, showing there are practical alternative ways of trading that can be more beneficial for the poor" (Chandler 2006, 256). The concluding chapter is about extending Fair Trade's reach. It documents Fair Trade's growth and current efforts to certify more products as well as to use the products in different ways, for example, as ingredients in prepared foods and body-care products. Chapter 7 goes on to address tensions that have emerged as Fair Trade has gone from being a niche offering for conscious consumers to—in some cases—a very mainstream offering of large, wholly profit-driven businesses. Is mainstreaming "watering down" the Fair Trade movement? I argue that it is not, as long as standards are upheld and producers are able to make informed decisions about which firms they will do business with. The political/transformative and market-centered components of Fair Trade must not become competing factions but must instead join efforts to alleviate poverty in the short term and change norms (from Fair Trade to fair trade) in the long term.

Chapter 1

Fair Trade from the Ground Up

There are over eight hundred Fair Trade certified producer organizations on four continents: Africa, Asia, North America, and South America.[1] Nearly half of them grow coffee; others produce tea, cocoa, grains, sugar, spices, fruit, honey, wine, olive oil, and flowers. Why and how have these producers organized? How do they perceive their participation in Fair Trade? What benefits and drawbacks have they experienced? How are they differentiating their products in the market? A body of field research by sociologists, anthropologists, geographers, and others allows us to address these questions. The studies that inform this chapter are listed in table 1.1.

WHY ORGANIZE?

Some producers were organized prior to their involvement in Fair Trade. In El Salvador, Mexico, and Nicaragua, coffee-growing cooperatives date back to 1980s land reforms. In their study of the Mexican cooperative union Coordinadora Estatal Productores Café Organico (CEPCO), researchers Josefina Aranda and Carmen Morales (2002) note that it was easy to achieve Fair Trade certification because the group and its member cooperatives were already organized and democratic. Similarly, in Tanzania the socialist government (in power from 1967 to the 1980s) assigned all coffee growers to rural cooperative societies that in turn aggregated into cooperative unions; some coffee unions there have even deeper roots. In post-apartheid South Africa, the policy of Black Economic Empowerment (BEE) has provided incentives to establish black-owned businesses as well as enterprises in which the workers are shareholders. These include vineyards, wineries, tea cooperatives, and fruit farms. Start-up funding for one Fair Trade winery came from the country's Department of Land Affairs and from a trust established to allocate public and government donations received after a disastrous bridge accident in 1987.

The first producer group in the world to gain Fair Trade certification basically invented a label that stated what they were already doing. The Unión de Comunidades Indígenas de la Región del Istmo (UCIRI) in Oaxaca was organized with the help of farmer/priest Franz (Francisco) VanderHoff Boersma, who then worked with a Dutch development organization to create a certification process and establish Max Havelaar, the first Fair Trade certifier (Audebrand and Pauchant 2009, VanderHoff Boersma 2009). Other Fair Trade producer groups formed with the aid of religious organizations. In Bolivia, the Central Cooperativas Agropecuarias "Operacion Tierra" (CECAOT) union of quinoa growers came together to mechanize their production at the urging of Belgian missionaries (Cáceres, Carimentran, and Wilkinson 2007). The Oaxaca, Mexico, cooperative Yeni Naván/Michiza and the Guatemalan highlands cooperative La Voz have histories that link Christianity, social justice, and a desire for autonomy (Jaffee 2007, Lyon 2002). Describing the origins of the La Selva coffee cooperative in Chiapas, Mexico, researcher Alma Amalia González Cabañas writes,

> Many of the communities where founding members live supported the work of consciousness-raising in the Catholic Church. The feeling of fraternity shared in the community reflection groups focusing on the "word of God" served as the basis for formation of relationships of trust (*confianza*) among the producers. The priests emphasized to the producers the importance of not relying on middle men (*coyotes*) to commercialize their products. They also promoted community projects in health care, basic supplies stores, literacy and, of course, the formation of catechists. (2002, 3)

Table 1.1
Field Studies about Fair Trade Producer Groups

Study	Product	Country	Producer Group(s)
Aranda and Morales 2002	coffee	Mexico	CEPCO*
Bacon et al. 2008	coffee	Nicaragua	CECOCAFEN,* PRODECOOP*
Besky 2010	tea	India	Darjeeling tea plantations
Cáceres and Wilkinson 2007	quinoa	Bolivia	CECAOT*
Doherty and Tranchell 2005	cocoa	Ghana	Kuapa Kokoo
Dolan 2008	tea	Kenya	Kiegoi
Frundt 2009	bananas	Dominican Republic	Finca 6 (also unions, plantations)
Frundt 2009	bananas	Windward Islands	WINFA*
Frundt 2009	bananas	Ecuador, Peru	El Guabo
Frundt 2009	bananas	Peru	El Prieto, CEPIBO
Garza and Trejo 2002	coffee	Mexico	Unión Majomut*
Getz and Schreck 2006	bananas	Dominican Republic	Finca 6
González Cabañas 2002	coffee	Mexico	La Selva
Jaffe 2007	coffee	Mexico	Yeni Naván/Michiza
Kleine 2008	wine	Chile	Los Robles
Lewis and Runsten 2008	coffee	Mexico	La 21 de Septiembre, Yeni Naván/Michiza
Linton 2008	coffee	El Salvador	APECAFE*
Linton 2008	coffee	Tanzania	AKSGC*
Lyon 2002	coffee	Guatemala	La Voz
Lyon 2007	coffee	Guatemala	La Voz
Lyon 2010	coffee	Guatemala	La Voz
Méndez 2002	coffee	El Salvador	APECAFE,* Las Colinas, El Sincuyo

Study	Product	Country	Producer Group(s)
Moberg 2005	bananas	Eastern Caribbean	WINFA (2nd-tier)
Moberg 2010	bananas	Eastern Caribbean	WINFA (2nd-tier)
Moseley 2008	wine	South Africa	Citrusdal, Thandi
Pirotte et al. 2006	coffee	Nicaragua	CECOCAFEN,* PRODECOOP*
Pirotte et al. 2006	coffee	Tanzania	KCU,* KNCU*
Raynolds and Ngcwangu 2010	tea	South Africa	Heiveld, Wuppertal
Ronchi 2002	coffee	Costa Rica	Coocafé*
Satgar and Williams 2008	coffee	Ethiopia	Oromia*
Satgar and Williams 2008	coffee	Tanzania	KCU*
Satgar and Williams 2008	coffee	Tanzania	KNCU*
Satgar and Williams 2008	tea	South Africa	Heiveld
Schreck 2005	bananas	Dominican Republic	Finca 6
Shuman et al 2009	cocoa	Ghana	Kuapa Kokoo
Sick 2008	coffee	Costa Rica	Coopeagri, Montes de Oro, CoopeSarapiquí**
Simpson and Rapone 2000	coffee	Mexico	UCIRI*
Utting 2009	coffee	Nicaragua	Soppexcca*
Utting-Chamorro 2007	coffee	Nicaragua	Soppexcca*, CECOCAFEN*
VanderHoff Boersma 2002	coffee	Mexico	UCIRI*
Walsh 2004	coffee	Peru	CEPICAFE*

* Second-tier organizations.
** Coocafé members.
Articles that include more than one case are listed multiple times.

Still other groups of farmers organized specifically because they wanted Fair Trade certification. This is true of the coffee unions Coocafé in Costa Rica, Oromia in Ethiopia, and CEPICAFE in Peru (Ronchi 2002, Satgar and Williams 2008, Walsh 2004). Once established, many groups receive start-up or goal-specific aid from nongovernmental organizations (NGOs). At least five different NGOs have supported "producer-led efforts to build effective cooperatives" in Nicaragua (Bacon et al. 2008, 261), and NGOs played an important role in financing Coocafé's *hijos del campo* rural education program and the group's production of the roasted and packed Café Paz brand (Ronchi 2002). South Africa's Heiveld and Wuppertal rooibos tea cooperatives achieved Fair Trade certification with the help of two NGOs (Raynolds and Ngcwangu 2010). Working with United States Agency for International Development (USAID) funds, the "business solutions" NGO TechoServe helped groups of Tanzanian farmers to improve their coffee's quality and form a new exporting entity, the Association of Kilimanjaro Specialty Coffee Growers (AKSCG). Subsequently, AKSCG's farmers applied for and received Fair Trade certification because they already met the criteria and wanted to expand their market reach (Linton 2008).

At least one Fair Trade producer group, Kuapa Kokoo in Ghana, exists in response to an International Monetary Fund/World Bank structural adjustment program. The Ghanaian government's denationalization of the cocoa trade made farmers more vulnerable but also allowed them to set up their own organizations. The UK-based Fair Trade organization Twin Trading provided start-up credit and financial advice to the initial woman-led group of 2000 cocoa farmers. "This was a farmer-rooted response to liberalization aimed at increasing farmer power and representation within the market as well as enhancing women's participation and encouraging environmentally sustainable production patterns" (*Report on Kuapa Kokoo Farmers 2001*, cited by Doherty and Tranchell 2005, 170).

Since the late 1970s, the Eastern Caribbean has been part of the Mabouya Valley Development Project (MVDP), which is aimed at developing the agricultural sector by way of individualized (rather than family) land tenure. In the early years of the MVDP, farmers could get low-interest loans to buy five-acre parcels from the government. They received subsidized irrigation and participated in local community organizations, with women in the forefront. By holding monthly meetings, MVDP promoted common interests and familiarity. The farmers organized into cooperatives because of the impending termination of tariff-free quotas in European markets.[2] Facing competition from giant Latin American plantations, in 1999 the Windward Islands Farmers Association (WINFA) started working with Fairtrade Labelling Organization (FLO; now Fairtrade International) to establish Fair Trade producers on St. Vincent, Dominica, St. Lucia, and Grenada. Before then, formal cooperatives "either did not exist or were largely nonfunctioning," so their proliferation can be attributed entirely to incentives provided by Fair Trade (Moberg 2005, 12). The first farmers to participate in Fair Trade were those who had previously participated in the MVDP. These farmers were more established and owned land, and the women had previous experience participating in mutual-aid groups (Moberg 2010).

These examples show very diverse motivations and circumstances for the acquisition of Fair Trade certification and for participation in Fair Trade value chains by various groups of farmers. Some cooperatives already existed as a result of land reform, faith- or NGO-based development initiatives, and response to structural adjustment or trade liberalization. These groups later applied for Fair Trade certification, whereas others organized specifically because they wanted the certification.

PRODUCERS' PERCEPTIONS OF FAIR TRADE

Field researchers usually spend a large amount of time living within Fair Trade producer communities and interviewing or surveying community members, so their field studies highlight a diversity of views and understandings on the part of organized farmers and their groups' leaders. Although it is impossible to definitively label Fair Trade producers' status within their groups, it is useful to differentiate between farmer-members, farmer-leaders, and union leaders. Farmer-members are, first and foremost, farmers. They do not claim leadership roles in their groups. Farmer-leaders organize, educate, direct, and in some cases act as the interface between their producer group and Fair Trade buyers. Union leaders work in offices; they are administrators of large groups of cooperatives that export produce together. The cooperatives in a union are not necessarily in the same geographic area. For example, Manos Campesinos in Guatemala is a union of eight cooperatives in four distinct regions of the country.

Farmer-Members

Participant observation indicates that the relationship between farmers and Fair Trade's goals is often ambiguous. For example, one study indicates that the farmer-members of a Mexican coffee cooperative saw poverty as part of their identity, believing that they needed to stay poor in order to benefit from Fair Trade (González Cabañas 2002). Others document Kenyan tea farmers' cluelessness about where their Fair Trade teas end up (Dolan 2008)[3] and Darjeeling tea plantation workers' scant knowledge of their rights under the Fair Trade system (Besky 2010). Out of fifty-three members of a Guatemalan coffee cooperative surveyed between 2001 and 2006, only three were even familiar with the term *comercio justo* ("Fair Trade"; Lyon 2007, 257).

A study conducted in a Peruvian coffee cooperative found that, although the leaders of the cooperative were very clear about who was buying the coffee, the prices paid, and the expenditures, farmers "find it difficult to comprehend exactly why different buyers pay more for the same product and why prices are higher in some years than others. . . . What seemed to stick in the end is that [the cooperative] exports directly to a range of international buyers who pay different prices and value different things. The common denominator is that all of these foreign buyers demand high quality. . . . Instead of attributing favorable prices to Fair Trade, producers tend to associate them with *membership in an organization that supplies high quality coffee directly to international markets*" (Walsh 2004, 39–40, italics added). This *is* something Fair Trade cooperatives do; but note that these producers saw Fair Trade as just another market.

Another study of coffee growers, this time in Oaxaca, Mexico, reveals that farmers relate to Fair Trade on the basis of price, services provided, or organic production, but not necessarily in terms of solidarity, alternative markets, empowerment, or sustainability. "Members are far more likely to identify themselves as organic coffee producers than as fair-trade producers, because it is the tasks involved in organic coffee production that actually influence their farming practices and differentiate them most clearly from their neighbors" (Jaffee 2007, 91). The Fair Trade market has provided these farmers with crucial price stability, but they "hold widely different understandings of what fair trade is" (Jaffee 2007, 91). Some equate "Fair Trade" with "organic"; others mention better prices or greater demands on the producers. In summary, sociologist Daniel Jaffee writes, "fair trade constitutes a critical—*yet largely invisible*—factor that stabilizes the much higher prices they receive for their organic coffee" (2007, 92, italics added). The Fair Trade–organic producers he studied are doing better than their neighbor-farmers who are not in the cooperative, but they still have trouble making ends meet. Given the higher standards and extra work that cooperative membership entails, Jaffee asked farmers why they stick with it. This is the part of the story where the less-tangible benefits of Fair Trade come into play. "For many producers, the main reason for sticking to the organization may be something impossible to quantify: the sense of belonging to something larger than oneself. . . . Those who have stuck with Michiza for several years do so because it makes them actors in a collective process, one that has deeper meaning than any simple measure of loss and profit" (Jaffee 2007, 245). Michiza members take pride in their organic farming, even as they struggle financially. Speaking about why farmers stay in the cooperative, one longstanding member told Jaffe,

> They get to know more people, they know that their coffee is being sold for a certain price, and they know where it ends up, and that nobody—at least not within the cooperative—is lining their pockets with the difference. They have more information, they have come to value many things that before they didn't value. They have discovered important things that the organization has offered them, such as how to use their local resources, and they enter into another kind of dynamic. And they say "Well, that's a profit too, right?" That's a profit. (245)

Research among producers in the nine Costa Rican cooperatives that comprise the Coocafé coffee union indicates that they understand Fair Trade largely as a price-stabilization mechanism. They appreciate having a fixed schedule of deliveries to Fair Trade clients rather than one-off orders. Also, belonging to a larger organization (the union) has improved the leadership within the cooperatives, which has led to better service delivery to farmer-members and the surrounding communities. Producers are very aware of the superior price conditions and the improved services of their cooperative, "but with a low awareness of Fair Trade" (Ronchi 2002, 25). Because Fair Trade has successfully provided a better price and supported the Coocafé union, it appears that farmers' lack of awareness of the source of their prosperity is a problem of communication that the cooperative leaders could address.

Interviewing banana growers in the Dominican Republic, sociologist Aimee Shreck found that all of the farmers correctly identified themselves as organic producers. More than three-quarters of them were also members of Fair Trade cooperatives, but only half of these recognized as much.

> When asked more specifically about the Fair Trade initiative (such as about its benefits and how it worked), even these producers demonstrated only an elementary and partial understanding, at best. For example, when asked about Fair Trade, producers with some knowledge often mentioned that they knew there was something called "comercio justo" (Fair Trade), but were unable to explain more. One producer's response captured a sentiment I frequently heard. To paraphrase his explanation: "Max Havelaar (the name of the Fair Trade organization that originally worked with these farmers) is a guy from Europe and he likes us small farmers, and so he buys our bananas." Others more concisely reported that Fair Trade was "a market" but could not elaborate. Significantly, none of the producers in my sample mentioned or knew about the minimum prices guaranteed by FLO, nor of the long-term commitment Fair Trade partners are expected to make to producers. (2005, 22)

The Eastern Caribbean banana growers Mark Moberg studied seem most familiar with the parts of Fair Trade that they do not agree with, especially Fairtrade International's prohibition of herbicides. Although admitting that better prices and the Fair Trade social premium have offset the costs of compliance to this rule, "farmers see little difference between the new standards to which they are held by FLO and the previous dictates of [exporters] and European importers to which they continue to be subject, all of which they characterize as arbitrary, costly and authoritarian" (2010, 67).

At the other end of the spectrum are studies that point to Fair Trade producers who thoroughly understand Fair Trade value chains and see themselves as part of a global social-justice oriented community, or a "solidarity-seeking commodity culture" (Bryant and Goodman 2004, 360). For example, Kuapa Kokoo farmers in Ghana are part owners of Day Chocolate, the company that buys most of their produce (Doherty and Tranchell 2005). The farmers, who are organized within 1,300 village-based societies, are stakeholders who control what they produce and sell (Shuman, Barron, and Wasserman 2009). Members of the Chilean Los Robles wine cooperative have a very deep understanding of Fair Trade as a social marketing strategy that differentiates them from their competition (Kleine 2008). VanderHoff Boersma's (2002, 2009) writings about Unión de Comunidades Indígenas de la Región del Istmo (UCIRI) emphasize farmers' solidarity, connection to the land, and maintenance of indigenous cultural practices even more than their improved standard of living. In some Nicaraguan coffee unions, farmers linked Fair Trade with high-quality and organic production, and recognized a need for solidarity *in order to meet these standards* (Pirotte, Pleyers, and Poncelet 2006).

In the research discussed above, the most common way that farmers relate to Fair Trade is in terms of stable or better prices and access to markets. Next we find ideas about social justice, solidarity, and—in one case—social marketing. Two of the studies reveal that many members of Fair Trade cooperatives do not know what "Fair Trade" means. And, finally, the Kenyan tea farmers did not know where their tea was going, but they were aware of development projects funded by the Fair Trade social premium (Dolan 2008).

Farmer-Leaders and Union Leaders

Farmer-leaders are at the helm of cooperatives or large individual farms; union leaders manage groups of producer cooperatives. These people interact with Fairtrade International and international buyers; they certainly need to understand their jobs. Independent cooperatives, farms (e.g., tea estates and vineyards), and unions all sometimes find it necessary to hire leaders (or at least consultants) from outside their membership to deal with export transactions, recordkeeping, and sometimes marketing as well.[4]

Researchers Víctor Pérezgrovas Garza and Edith Cervantes Trejo's study is especially valuable here because it explicitly differentiates between farmer-leaders and union leaders within a particular union. Cooperative leaders who had listened to union leaders talk about the Fair Trade system still "do not have a clear idea of what the 'Fair Trade system' is; rather they identify Fair Trade with the buyers who have come to visit the cooperative." The union's leaders and technical team, however, exhibited "a fairly precise understanding of the Fair Trade system . . . who the different actors are in the system: importers, producers, inspectors and distributors. They clearly understand the minimum guaranteed price, the social and organic premiums, and the system of pre-financing" (2002, 15).

Unions of cooperatives face the challenge of maintaining effective communication with leaders of the often far-flung groups under their umbrellas, and some of them are meeting this challenge. In her most recent visit to the Nicaraguan Soppexcca coffee union, environmental social scientist Karla Utting (2009) noticed that farmers' (and, by extension, farmer-leaders') understanding of Fair Trade had increased. This was due at least in part to the union leaders' use of radio broadcasts. A comparative study of African cooperatives indicates that education about Fair Trade principles is quite effective *when combined with instruction in techniques to improve productivity and/or organic practices* (Satgar and Williams 2008).

People in leadership roles generally comprehend more about Fair Trade values than rank-and-file farmer-members. Some groups were built on Fair Trade principles to begin with; their leaders pass on values they have inherited. For others, face-to-face interaction with Fair Trade buyers or labelers has changed the way they perceive what they are doing. Within unions, top-level leaders can do much to educate group-level leaders. Not surprisingly, the most effective educational outreach combines practical topics (e.g., how to increase productivity) with more socially or environmentally oriented messages.

Regardless of producers' knowledge of Fair Trade principles or international networks, they readily enumerate the benefits and drawbacks that have accompanied their experiences. The same farmers who said that they needed to stay poor in order to participate in Fair Trade also acknowledged that Fair Trade "strengthens organizations and permits learning about how the market functions in a less competitive atmosphere" (González Cabañas 2002, 33). Fair Trade producers commonly cite better prices and technical assistance as benefits of Fair Trade. Some specifically mention credit, being able to afford better houses and education for their children, and simply being able to stay on their land (see, e.g., Jaffee 2007, Méndez 2002, Ronchi 2002). In groups whose members associate Fair Trade with solidarity and social justice, farmers are likely to highlight benefits, such as cultural preservation, increased democracy, women's empowerment, community development, or a way to be heard by governments (see, e.g., Kleine 2008; Lyon 2002, 2007; Simpson and Rapone 2000; Utting 2009; VanderHoff

Boersma 2009). Very frequently, the Fair Trade advantages producers perceive derive specifically from their groups' use of the Fair Trade social premium, for example, improved health care, education, infrastructure, and processing/shipping facilities for their produce.

ONGOING CHALLENGES

Achieving Fair Trade certification is a formidable process involving inspections, additional record keeping, and fees; yet acquiring certification does not guarantee that a producer group will sell all (or any) of its produce under Fair Trade terms. Low Fair Trade sales may be due to a group's newness to Fair Trade markets or to inadequate quality. These matters can usually be addressed over time, but quality improvements require monetary investment in facilities and training. Country-level NGOs and Fair Trade labelers have been helpful in providing loans and training, as have international NGOs such as Oxfam. Still, field studies show that many producer groups and their members confront challenges in regard to their participation in Fair Trade. Below I discuss the three most notable of these: paying certification fees, maintaining rural communities, and empowering women.

Certification Fees

As with organic certification, Fair Trade certification is not free. For example, groups of fifty or fewer producers pay a €500 (about $680) application fee (€150 for each additional product) and €400 (about $545) per day for initial certification services. Annual recertification fees are based on the rate of €350 (about $475) per day. Fairtrade International offers grants that cover as much as 75 percent of the fee, but how much applicants actually receive depends on the demand for grants each year.[5] Field studies often record complaints about certification fees, but none of them note whether or not a producer group applied for or received a grant.

Maintaining Rural Livelihoods

Fair Trade producers find themselves better off than their independent small-farmer neighbors or plantation workers, but some of them still face food insecurity (Bacon et al. 2008). Fair Trade does not impede networks of migration from Latin America to the United States, although it does prevent or slow within-country rural-to-urban migration (Ronchi 2002, Utting-Chamorro 2005, cf. Mutersbaugh 2002). Increased crop diversification and government interventions can allay food shortages, but entrenched networks of international migration still perpetuate themselves. In many Mexican coffee-growing locales, international migration is embedded as a practice of household-level economic diversification and risk management—what social scientists call the "new household economics of migration" (Massey et al. 1993). Fair Trade brings a better price, but farmers need to hire workers rather than rely on the labor of family members who are in the United States. This significantly adds to the cost of production and reduces the economic viability of growing coffee—even Fair Trade coffee (Jaffe 2007, Lewis and Runsten 2008).

Also of note is that the full Fair Trade price does not always get to farmers because their cooperatives use some of these revenues to service debts or pay administrators' salaries (Jaffee 2007, Mohan 2010). If cooperatives or unions are in debt, the prices that individual farmers actually receive can be substantially lower than the Fair Trade price because the producer group must service its creditors before paying its members. For this reason, Jaffe recommends altering the Fair Trade standards to requiring a "farm-gate price" paid directly to farmers rather than to their cooperatives or unions. The situation is potentially graver for workers on Fair Trade certified plantations. For example, anthropologist Sarah Besky studied plantation workers in West Bengal, India, where high grades of Darjeeling tea fetch attractive prices. She found that Fair Trade certification was benefitting plantation owners, not workers, because the latter received the same legally mandated wages regardless of how much the tea is sold for. Documenting owners' use of the Fair Trade social premium to pay for things the government requires them to provide for their workers anyway, such as housing and food, Besky concluded, "In Darjeeling, Fair Trade standards . . . have undermined what local laws and government labor officers have done to promote social justice" (2010, 99). Furthermore, some buyers were not even paying the social premium. They market the "tea from a Fair Trade plantation" as opposed to "Fair Trade tea"—a distinction lost on all but the most discriminating consumer.[6]

Empowering Women

Evoking an ancient Chinese proverb, journalists Nicholas Kristoff and Cheryl WuDunn (2009) remind us that women hold up "half the sky," and argue quite convincingly that women's rights are a (if not *the*) paramount global issue of this century. How can any society prosper if half of its members (the half that takes on the lion's share of raising children) do not benefit from the same education, health care, freedoms, and opportunities as the other half? Several studies evaluate women's empowerment and involvement with Fair Trade producers. In Nicaragua, agroecologist Christopher Bacon and his colleagues (2008) found women to be heavily involved in the work of coffee farming, but this did not lead to greater empowerment within their families. Other studies note the near absence of women in leadership roles. For example, although nearly 20 percent of a prominent Costa Rican coffee union's members are women, their representation is negligible. Domestic responsibilities often prevent them from attending meetings or actively participating in their cooperatives' governance (Ronchi 2002). Sarah Lyon's study of women in a prominent Guatemalan coffee cooperative reveals their exclusion from the group's leadership and subsequent attempts to create "a market of their own" for their weaving. These women are "not pursuing an abstract notion of gender equity that mirrors the concerns of Northern consumers." Rather, they are seeking to improve their livelihoods in a way that is appropriate in their culture (2010, 143).

By contrast, in Michiza, the Oaxacan cooperative that Jaffee (2007) studied, participation in Fair Trade has helped women to become leaders. The difference may have to do with the fact that emigration to urban centers and the United States has severely reduced the male population in many Oaxacan villages, including Jaffee's site. Traditional norms that gender various coffee production tasks have become blurry. When women are doing "men's" work, such as plot maintenance, or are in charge of farms in male household members' absence, the men who remain cannot (and do not even try to) justify women's exclusion from the cooperative's leadership. The Heiveld tea cooperative in South Africa is a more progressive example of linking Fair Trade and women's empowerment. There, *individuals* (rather than families or farms) comprise the cooperative's membership (Satgar and Williams 2008).

DIFFERENTIATED QUALITY

How do producers get a foothold in Fair Trade value chains? To answer this question, we have to understand the relevant markets. Fair Trade is a "different type of market" but it is not "charity trade" (VanderHoff Boersma 2009). Companies and end consumers may wish for their sourcing and buying decisions to convey a global social consciousness, but in doing so they expect quality commensurate with the price. Businesses whose existence depends on the quality-differentiated products they purvey (e.g., specialty coffee roasters, wine merchants, purveyors of fine teas, or *chocolatiers*) literally cannot afford to buy Fair Trade products unless these products meet particular standards. And businesses selling fairly undifferentiated products, such as bananas, need a strategy for linking the Fair Trade message to the fruit.

Coffee

Most Fair Trade coffee is *Coffea arabica*, the tasty species that typically grows at high altitudes. Cup quality reflects growing and harvesting conditions (e.g., altitude, soil quality, weather) as well as the way the coffee cherries are processed to yield green (unroasted) coffee beans for market. Writing in 2001, the specialties editor of the trade journal *Tea and Coffee* remarked that "a guaranteed premium price without a guaranteed premium cup is not sustainable." He saw Fair Trade as a laudable movement, but one that did not always provide laudable coffee (Schoenholt 2001). Since then, much emphasis has been placed on quality because Fair Trade coffee sells mostly within specialty markets, where customers' choices are motivated more by taste characteristics than by price.[7] Fair Trade labelers and other NGOs have stepped up their efforts to help growers and potential buyers learn about each others' needs and constraints by arranging for coffee industry representatives' visits to farmer cooperatives and sponsoring cooperative representatives' visits to specialty coffee trade shows.

Most coffee-specific field studies explicitly discuss quality. They mention quality-improvement practices, such as hand picking and sorting, washed processing, and tasting samples of roasted coffee in cupping laboratories. Researchers Jessa Lewis and David Runsten's (2008) comparison of Oaxacan coffee growers within and outside of Fair Trade cooperatives is particularly interesting because it shows how being organized has given cooperative members better access to training and high-quality processing facilities. But several studies also note that, especially in sub-Saharan Africa, infrastructure limitations prevent organized farmers from processing much of their coffee for the specialty export market (Linton 2008; Pirotte, Pleyers, and Poncelet 2006; Satgar and Williams 2008). Meanwhile, some low-altitude cooperatives have attained Fair Trade certification but are still not able to sell to Fair Trade buyers because their coffee does not meet a differentiated standard, regardless of how it is processed (Sick 2008).[8]

Wine

The few field researchers studying Fair Trade vineyards and wineries seem to take quality relative to price as a given. Geographers William Moseley (2007, 2008) and Dorothea Kleine (2008) barely discuss it in their articles about Fair Trade vineyards and wineries in South Africa and Chile, respectively. But, as was true of coffee, the initial offerings of Fair Trade wine were uneven. UK-based wine buyer and consultant Angela Mount recently told the *Guardian*, "A few years ago, to be honest, the wines just weren't up to it," citing producers' inexperience and initial prioritizing of quantity over quality as contributing factors. But now some Fair Trade wines offer "extremely good value" (quoted in McEvedy 2009).

Quality and value are behind wine's rise to become one of the fastest-growing Fair Trade products in the United Kingdom, where consumers can choose from many varieties and labels (Weekes 2008). Fair Trade wines are becoming more and more obtainable throughout Europe, and of late one can buy excellent and competitively priced Fair Trade wines from South Africa, Chile, and Argentina in the United States. South African Fair Trade wines are starting to win quality awards and four-star ratings in the country's wine bible, *John Platter's South African Wine Guide*. The Thandi Fair Trade winery in the Elgin region of the Western Cape province even produces a formidable pinot noir, although this grape is risky to grow in such a sunny climate.

Tea

Little has been written about Fair Trade tea, let alone about its quality or markets. Writer Beatrice Hohenegger's (2006) *Liquid Jade: The Story of Tea from East to West* illuminates the beverage's historical and cultural significance and details labor and environmental concerns around contemporary tea farming, but only mentions Fair Trade in passing. Anthropologist Catherine Dolan conducted research at the Kiegoi tea factory in Kenya (one of eight Fair Trade–certified groups exporting via the Kenya Tea Development Agency Limited). This tea's quality is so high that farmers sometime receive more than the Fair Trade price. Still, some of Kiegoi's tea sold via Fair Trade channels ends up in blended "Fairtrade tea" . . . of several different teas. This "provides buyers with considerable latitude in sourcing locations" (2008, 308).

Of Fair Trade USA's twenty-two Fair Trade tea producer group profiles, nine mention a distinguished, location-specific appellation (e.g., Nilgiri, Assam, Darjeeling), seven are 100 percent organic, and three engage in biodynamic practices.[9] These quality-indicating markers overlap considerably; about half of the groups have *none* of them. Since the early 2000s Fairtrade International has been careful to link certification to market demand—only certifying producer groups and new products when a demonstrated market exists, for example, a buyer requests that a group attain Fair Trade certification, or a food manufacturer asks for Fair Trade ingredients such as spices and nuts. The fact that half of the tea producers profiled by Fair Trade USA make no claims about differentiated quality suggests that there is a Fair Trade market for low-priced tea bags as well as for high-end, appellation-specific teas. Shopping for Fair Trade tea in the southwestern United States, I find inexpensive boxes of tea bags made from low-quality fannings (the small particles left over after tea leaves are processed) and a good selection of higher-quality, higher-cost bagged options. For premium loose teas (the best quality *and* best value), I need to shop online. Loose tea is a very small (albeit rapidly growing) segment of the US tea market, but the picture is quite different in other countries. Globally conscious tea drinkers in the United Kingdom, Germany, and Japan can choose from a broad range of premium loose-leaf Fair Trade varieties in their local shops.

Chocolate

"Made with love, integrity, and only the highest quality, sustainably sourced ingredients" reads the wrapper of an organic, Fair Trade, 70 percent cacao cherry and almond chocolate bar I recently purchased. Like coffee, cacao beans' quality is differentiated by growing altitude, climate, and processing. Fair Trade cocoa farmers have successfully tapped into gourmet chocolate markets, selling to small companies that offer unique products at premium prices. In the meantime, campaigns to get large US candy makers, such as Hershey's, Mars, and Nestlé to source Fair Trade cocoa—including an effort involving child lobbyists at *Charlie and the Chocolate Factory* movie-release events (Global Exchange 2007) and "reverse trick-or-treating" (Global Exchange 2010)—have been somewhat disappointing in their effect. But in the United Kingdom, the thriving Day Chocolate Company (co-owned by Twin Trading; Kuapa Kokoo; and The Body Shop, which uses Kuapa Kokoo cocoa butter) gained market share by emphasizing their mainstream Divine and Dubble brands at mainstream prices. Reviewing the Day collaboration, social enterprise scholar Bob Doherty and Sophi Tranchell, managing director of Divine Chocolate, note that Divine's "bean to bar" story about Fair Trade was more winning in the marketplace than higher-end Green and Black's emphasis on their (Fair Trade) chocolates' quality and uniqueness (2005). Since autumn 2009, venerable UK-based Cadbury has used Fair Trade cacao beans in all of their milk chocolate (Cadbury 2009).[10]

Why, in the United States, is Fair Trade cacao the main ingredient in high-end, special-occasion chocolate, whereas in the United Kingdom it easy to find it in good-quality but largely mainstream, "everyday" candy bars? Part of the answer lies in the fact that the major US chocolate makers are enormous compared to their UK counterparts; the US companies see commitments to Fair Trade sourcing as an unrealistic constraint on their supply chains. But there's more to the story: Britons are savvier to global trade issues, *and they eat a lot more chocolate.* On average, consumers in the United Kingdom eat thirty-five pounds per year, compared to about twelve pounds in the United States (Doherty and Tranchell 2005, GourmetSpot 2009 [statistics from Hershey's]). This combination of consciousness and consumption suggests that Fair Trade chocolate is a much more pressing concern in the United Kingdom.

Bananas

Bananas are the world's most popular fruit. Unlike unroasted coffee, processed cacao beans, and tea leaves, they (and other fruits that have recently come under the Fair Trade umbrella, such as mangoes) are highly perishable. Furthermore, bananas are largely undistinguished in the market; shoppers assess quality by appearance, not by taste. US supermarkets offer only one or perhaps two banana options (e.g., conventional and organic), sold loose. Fair Trade bananas first appeared in 2004, but at present it is very hard to find the Fair Trade label on a banana sold in America. Despite promising sales figures, "supply, shipping, ripening, and distribution problems together undermined the quality and quantity of Fair Trade bananas available in the United States" (Raynolds 2007, 73; see also Frundt 2009).[11]

The situation is quite different in Europe, where Fair Trade bananas have been on offer since 2000. Because the shift to Fair Trade was instigated by growers as a proactive response to the end of trade protections, it was not necessary to establish new shipping and distribution channels for Fair Trade fruit from the Caribbean. Fairtrade Germany's website lists three major supermarkets where Fair Trade bananas are available throughout the country. In the Netherlands, shoppers encounter Fair Trade bananas from six different countries at all of the country's major grocery chains. And some European marketers *do* differentiate bananas. For example, in the United Kingdom, "better" bananas are sold in packages that identify them as Fair Trade, organic, or kid-sized (with labels large enough to show more than a logo), in direct contrast to generic, unpackaged fruit (Moberg 2005).

CONCLUSION

Farmers in the global South are benefitting from Fair Trade—significantly or at least a little. There are diverse motives for and experiences of group formation, very disparate understandings of Fair Trade, and correspondingly varying degrees of social transformation associated with it. Product-specific strategies of market differentiation, however, are quite similar across producer groups. It is noteworthy that Fair Trade chocolate and tea are successfully penetrating medium-quality, medium-price markets as well as high-end ones. From Fair Trade producers' perspectives, there is evidence of economic and social benefits, successful problem solving, and ongoing dilemmas and challenges. The case studies discussed here help us think about ways to strengthen Fair Trade and inform its expansion.

Field researchers embark on their projects with specific goals, disciplinary backgrounds, strengths, and limitations. The findings they publish are quite disparate—a positive in that the studies do not all revolve around a single set group of questions to the exclusion of other potentially relevant ones. This diversity, however, makes it difficult to systematically compare studies.[12] Below I present three promising directions for future research.

An Integrated Model

Independent consultant Karla Utting (2009) has proposed and preliminarily tested a comprehensive template for evaluating "responsible trade initiatives." I do not suggest that every study should attempt to examine all of

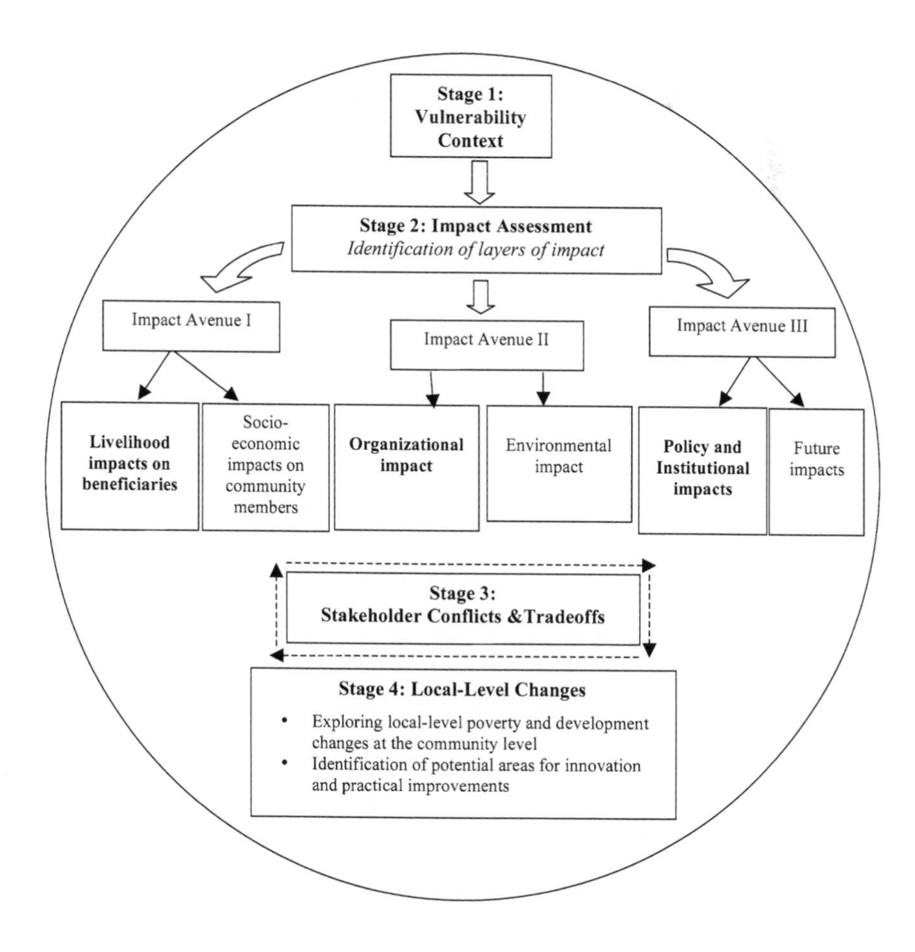

1.1 Impact Assessment Framework for Responsible Trade Initiatives
Source: Utting 2009

the elements in her model (which is particularly well suited for research in one locale over time) but rather that it guide future studies' lines of inquiry.

A primary strength of Utting's template is that it prioritizes context (stage 1), which could include country- and locality-specific factors as well as geographic, cultural, and market circumstances particular to growing a crop (coffee, tea, bananas, grapes, and so on). Then, in stage 2, producers adopt an "intervention" (e.g., Fair Trade). How does participation in the "intervention" affect farmers' livelihoods? Are there broader, community- or regional-level outcomes? Does organizing empower individuals and promote environmental stewardship? Are member-farmers and -leaders setting an example that has or could influence government policy, or incite government cooperation to achieve particular goals? What are the possibilities?

Stage 3 of Utting's model acknowledges that there will usually be conflicts and tradeoffs, and encourages producers, activists, and scholars to recognize and tolerate these rather than abandoning ship. Then stage 4 asks us to return to the local point of view and see not only what has been accomplished but what could be. Fair Trade certification is a main artery leading to the heart of responsible global trade, but what constitutes "responsible" is a progressive target.

Paying Attention to Outliers

With few exceptions, studies reveal quite a gap between producer group members' and leaders' understanding of Fair Trade. However, none of the studies consider why some farmer-members are atypical. What is different about the farmers who *do* convey an understanding of Fair Trade? And, since it is doubtful that every single member of Los Robles, UCIRI, or Kuapa Kokoo (the producer groups with more knowledgeable members) can articulate what Fair Trade is all about, what is different about the farmers who cannot? I take up the first question in the following chapter.

New Products, New Issues

Readers have probably noticed that many of the case studies discussed here are about coffee producers. Coffee still dominates Fair Trade markets, but as newly certified products gain market share we need more "not-coffee" studies. Certifying new Fair Trade products is a formidable effort. How should stakeholders determine a fair price? What are fair standards for hired labor? What environmental issues exist in this value chain, and which ones should Fair Trade standards prioritize? Consider grapes and wine, for example, which have only been Fair Trade certified since 2003. How do their value chains operate in various producing countries? Does the Fair Trade practice of linking grape-growing and wine-making operations in order to maximize the finished product's value invite unintended consequences? Might it be tempting to sell grapes to non–Fair Trade wineries for quick cash, or for Fair Trade wineries to buy grapes from other growers in order to enhance their blends or increase the variety of wines they have to offer? Under what circumstances might such practices be *compatible* with Fair Trade?

Drafting and revising Fair Trade guidelines is a process, not a circumscribed task. It will continue to involve industry-specific trade and marketing experts, but Fairtrade International and the labeling NGOs, informed by their dealings with producer groups and to some extent by the research presented here, have made it their business to listen to and address farmers' concerns, increasingly viewing Fair Trade standards as part of an inclusive and evolving process. The standards are reviewed, and often revised, at least every five years (FLO 2006).

The development of Fair Trade standards for South African fruit, wine, and rooibos tea is an example of what is achievable. In their engaging account of the process, researchers Sandra Kruger and Andries Du Toit show that it was possible to "reconstruct fairness" so that the standards would reinforce efforts to promote an equitable transformation of the country's industries. In response to initial—somewhat contradictory—concerns voiced by stakeholders within and outside South Africa, Fairtrade International shifted a forum it intended to hold in Bonn to Grabouw, in an important fruit-producing region of South Africa. This was the first time Fairtrade International engaged in a process of direct consultation with multiple stakeholders, including other NGOs, activists, farmers, workers, and members of the government. Kruger and Du Toit do not claim that everyone had an equal voice in these negotiations; in fact they point out situations in which specific individuals were heavily influential due to their knowledge and skills, and that usually representatives of NGOs did the talking for the workers. But the outcome was Fair Trade "conventions more sensitive to the interests of South African workers," and "a process that entrenched a more robust 'empowerment' agenda" (2007, 217). This example shows that it is possible to maintain a high and consistent standard while taking into consideration farmers' unique circumstances, a positive development indeed.

Chapter 2

Fair Trade Coffee in Guatemala

Guatemala is the world's eighth-largest coffee-exporting country. More than 80 percent of the country's coffee farms are in the hands of smallholders (defined as farmers or families with less than 10 hectares or 24.7 acres) who grow coffee under the shade of larger trees. The combined potential for improved livelihoods and the protection of globally important biodiversity has attracted NGOs and specialty coffee importers to initiate a number of sustainability-oriented projects. Of these, Fair Trade—introduced in Guatemala in 1991—is by far the largest and most well known, but its results have not been systematically studied.[1] How are participants and their communities benefitting from Fair Trade? How do farmers view their involvement in Fair Trade cooperatives? What lies ahead for them? These questions guided a survey- and interview-based study of leaders and members of Guatemalan Fair Trade coffee cooperatives and cooperative unions.[2] This research augments the existing literature by taking a closer look at how Fair Trade is working on the ground in Guatemala. It also pays attention to interviewees who do not "fit" general patterns. To the extent that we can explain *why* these individuals are different, we can say something about *how* farmers come to have a deep understanding of Fair Trade principles.

Almost all of the Fair Trade producer groups in Guatemala grow coffee.[3]

The internationally recognized Fair Trade coffee standards are as follows:

* Smallholder producers are organized in democratically governed cooperatives or groups thereof.
* Sellers and buyers establish long and stable relationships. If asked to do so, buyers must extend up to 60 percent credit on contracts.
* Buyers pay producers at least the minimum Fair Trade price, currently $1.40 per pound for high-grade washed Arabica coffees, $1.70 per pound if the coffee is also certified organic. When the market price is higher than the Fair Trade minimum, the market price applies.
* In addition, buyers pay a social premium of twenty cents per pound. At least five of the twenty cents must go towards improvements in productivity or quality; the rest of the premium is expended at the discretion of the cooperative/group (FLO 2011a).

PREVIOUS RESEARCH

Only one field study that has yielded published results precedes this research: anthropologist Sarah Lyon's (2002, 2006, 2007, 2010) ongoing participant observation and surveys in a cooperative of indigenous farmers near Lake Atitlán. Her work has yielded several excellent articles about the group's gains and challenges, relationships between Fair Trade producers and consumers, the Fair Trade–human rights link, and the limitations faced by women in the cooperative. Lyon acknowledges several ways in which her research site is not generally representative of Guatemalan Fair Trade coffee cooperatives: It has been Fair Trade certified since 1991, has long-standing ties to Fair Trade buyers (including one that brings tourist groups every year), has received a USAID loan and generous aid from NGOs and coffee roasters, and until recently sold all of its coffee to Fair Trade buyers.

Lyon's site is, however, very representative of the challenges that confront small farmer organizations in Guatemala. The country has a long history of "violent repression, structural inequality, and cultural discrimination against indigenous populations and community organizers" (Lyon 2007, 241) as well as general corruption and exploitation of the relatively powerless by the relatively powerful. Acquiring Fair Trade certification does not erase this history or remove incentives for corruption. The cooperative has been suspended from participation in Fair Trade because of massive "side-selling"—in one case selling eighteen containers (about 594,000 pounds) of green coffee already contracted to Fair Trade buyers outside of the system, shorting the rightful buyers of their coffee and of whatever prefinancing they had provided. In the meantime, $400,000 disappeared, and the president of the cooperative fled to the United States.

Despite these serious setbacks, Lyon documents ways in which participation in Fair Trade has raised incomes and otherwise benefitted this cooperative's members and neighbors. The cooperative has fostered "safe opportunities for members to work together and reproduce long-term traditions of horizontal cooperation, reciprocity, and mutual aid in the pursuit of local community development" (2007, 252). The longitudinal design of her research has allowed her to watch events unfold and produce dynamic, in-depth ethnographies—something the present study cannot do. However, comparing several cooperatives, as I do here, makes it possible to understand interviewees' statements within their different contexts and to produce a larger, although less detailed, picture.

A BROADER LOOK AT FAIR TRADE IN GUATEMALA

There are twenty-two Guatemalan Fair Trade–certified coffee producer groups in FLO-CERT's registry. Seven of these are second-tier organizations—unions of six- to thirty-seven-member cooperatives. The groups established themselves between 1966 and 2005 and, as of the time this research was conducted, had Fair Trade histories that ranged from as long as eighteen years to as short as one year. The youngest cooperative has yet to sell anything via Fair Trade channels; others sell Fair Trade coffee in as many as twelve countries. Twenty individual cooperatives or unions took part in an extensive survey conducted in January and February 2008.[4] We asked concrete questions about coffee production, Fair Trade sales, revenues, expenses, and uses of the social premium as well as open-ended questions about governance, transparency, and how individual producers understand Fair Trade. Data about four nonsurveyed cooperatives (included in tables 2.1 and 2.2) come from Fair Trade USA's producer profiles and from producer groups' own websites.

Eight indicators of Fair Trade's potential to improve conditions for farmers and their communities comprise table 2.1's descriptive snapshot.

Table 2.1
Descriptive Statistics, Guatemalan Fair Trade Coffee Cooperatives and Unions, 2008

	N*	Average	Range
No. of labelers group sells through	24	3.17	0–12
Union of cooperatives (1=yes)	24	0.29	0–1
Annual production (category)**	24	1.63	1–3
Years FT certified	24	8.38	1–18
Percent organic	22	46.20	0–100
Coffee has earned a quality award (1=yes)	20	0.35	0–1
Access to cupping facilities (1=yes)	20	0.80	0–1
Premiums help broader community (1=yes)	24	0.46	0–1

* Number of groups observed
** 1=<16, 2=16–92, 3=>92 containers

The number of labelers a group works with tells us in how many countries they are selling Fair Trade coffee. *Unions of cooperatives* are larger and often have better resources for quality assurance, improvement, marketing, and recordkeeping. *Annual production* is another gauge of size and potential; more Fair Trade coffee to sell means direct benefits for more farmers and their families and more social premium money to work with. *Years of Fair Trade certification*, a proxy for experience and connection to buyers, could also be related to how much of their coffee members of Fair Trade cooperatives are able to sell under Fair Trade terms. Because organic coffee fetches a higher price, a cooperative's *percentage of organic beans* is another gauge of its earning potential.

The next two factors designate coffee quality and a group's potential to improve quality. *Cup of Excellence awards* are the result of strict competition between the best coffee producers in a given country for a particular year.

An Internet auction facilitates the sale of winning coffees to the highest bidders (Cup of Excellence 2011). *Cupping facilities* equipped with miniature roasters, grinders, and brewing equipment allow cooperative leaders and members to assess taste characteristics in a way that visually inspecting the preroasted beans does not. The material quality of coffee is due to the genetic type of tree (*arabica* or *robusta* —most Fair Trade coffee is *arabica*); agro-climate conditions, such as altitude, soil conditions, and rainfall; farm practices; harvesting practices (picking only ripe cherries); primary processing; and handling and storage along the value chain. Since coffee drinking is not part of indigenous Guatemalan food culture (and even if it were, local ideals and indicators of quality could differ from those set by specialty coffee roasters in the global North), farmers relate to quality mostly through farm practices such as fertilizing, pruning, weeding, mulching, watering, and shade-growing (Daviron and Ponte 2005, 130). They need to know what global buyers expect, and whether or not they are meeting expectations. In her study of the Nicaraguan Fair Trade union Soppexcca, Karla Utting interprets farmers' interest in cupping their own coffee as evidence of capacity building within Fair Trade networks. Many of her interviewees

> stated their lack of interest in tasting their own product before their involvement with the fair trade system. This may be because they were not aware of the information that cupping could provide, or how that information could be valuable to them. Now, half of the interviewees claim to have professionally tasted their finished product by having access to Soppexcca's cupping laboratory. (2009, 10)

The first seven aspects discussed above could all be related to a Fair Trade producer group's financial success. The last one, *whether or not a group is spending some or all of the social premium in ways that benefit the community at large* (rather than only member-farmers and their families) gauges Fair Trade's broader role in rural development.[5]

Table 2.2 presents simple two-way correlations, with statistically significant relationships[6] in bold type. It shows that years of Fair Trade certification, annual production, and second-tier status are positively correlated with access to markets. Access to markets and years of Fair Trade certification are most strongly associated with the eighth variable: premiums directed toward projects that contribute to community development and

Table 2.2
Correlations (N=24)

	(1)	(2)	(3)	(4)	(5)
No. of labelers (1)	1.00	**0.46**	**0.44**	**0.67**	**-0.68**
Union of cooperatives (2)	**0.46**	1.00	**0.75**	0.15	-0.33
Annual production (3)	**0.44**	**0.75**	1.00	0.22	-0.21
Years certified (4)	**0.67**	0.15	0.22	1.00	**-0.56**
Premiums help broader community (5)	**0.68**	0.33	0.22	**0.56**	1.00

Note: Statistically significant correlations are in bold. Variables that are not significantly correlated with any other variables are omitted.

empowerment. In other words, older, larger groups are selling more under Fair Trade terms; this gives them more premiums, at least some of which go toward community-level projects.

How do Guatemalan Fair Trade groups make decisions? How do their members view and understand their participation in Fair Trade? Below, the answers to these queries are grouped according to the type of questionnaires (leader or farmer-member) used to guide the interviews.

Leaders

The questions we asked leaders of cooperatives and cooperative unions focused on democratic decision making, transparency, quality improvement, and the use of the Fair Trade social premium. Everyone told us that members of their group's governing committee are elected (rather than appointed), and in all but one case the cooperatives' books are open to members' perusal. Emphasis on quality control and improvement is likewise ubiquitous.

Regarding how to spend the premium, 29 percent (five of seventeen) of the cooperative leaders and 50 percent (two of four) of the union leaders said that the decision-making process was controversial.[7] All of the union leaders agreed that the premium-spending decision was controversial; they cited contention over how to use the premium *among farmer-members* (though not necessarily among the farmer-leaders who comprised the union's leadership).[8] For example, the leader of a union whose cooperatives has previously spent premiums on education, health care, farm improvements, and buying more property reported that cooperative leaders had recently voted to divide the premium among individual farmer-members. "The people want to see the cash." Another union leader said,

> Many times the people ask, "Where are the schools and the health clinics?" The point is that we work in a certain way with the social premium. They want to see what Fair Trade has done and it has to be something tangible. This is sometimes difficult. So many times we tell them that one of the major effects of Fair Trade is that the producers today are still producing coffee and they have not cut it down and done something else. What we have seen is that other communities where they did not do Fair Trade, they have either left coffee or they have had to sell their land and now they have much less land to cultivate. There has been migration to the cities and out of the country. So the members aren't always in the best conditions, but they are producing coffee, and they are in their own communities. Some have been able to buy more and increase their production; however, the most important thing is that they are working in their communities.

Table 2.3 shows what the Fair Trade social premium is supporting, regardless of whether spending decisions were reached via consensus or via controversy and compromise. Some or all of these funds are most frequently directed toward improving the quantity and quality of export-ready coffee that a group can produce. Four cooperatives have established credit unions (or similar) for members, and four are so new that they have not yet done anything with the premium. Spending on education is also a popular choice, be it to send members' children to school or to help make schools better.

How do leaders view their organizations' involvement in Fair Trade? In voluntary comments they expressed support for Fair Trade as a principle and as a way to gain recognition in global markets, but they also gave quite a bit of criticism. Some reported bad experiences with Fairtrade Interna-

Table 2.3
Use of the Fair Trade Premium as Reported by Cooperative or Union Leaders (N=20)

Expenditure	No. of Mentions
Improve coffee infrastructure/provide training	9
Credit to members	4
Nothing yet	4
Improve schools	3
Scholarships for members' children	3
Office for the cooperative	2
Build roads	2
Direct cash transfer to members	2
Buy coffee (prevent side-selling)	2
Buy land	1
Install electricity	1
Health care for members	1
Community health clinic	1
Economic diversification (women-led project)	1

Note: Most cooperatives spend accumulated premiums on more than one thing.

tional (e.g., inspectors not arriving as scheduled, never receiving inspection reports, not knowing who they were supposed to communicate with, e-mail written in German). There is widespread concern about the price of certification and the amount of bureaucracy involved in Fair Trade value chains; both add costs and raise barriers to entry and ongoing participation.

Throughout most of these groups' histories with Fair Trade, free market coffee prices were quite low (figure 2.1 shows yearly average prices for the mild *arabica* coffees grown in Guatemala). Now that prices are better, the benefits of Fair Trade do not seem as obvious. However, cooperative leaders

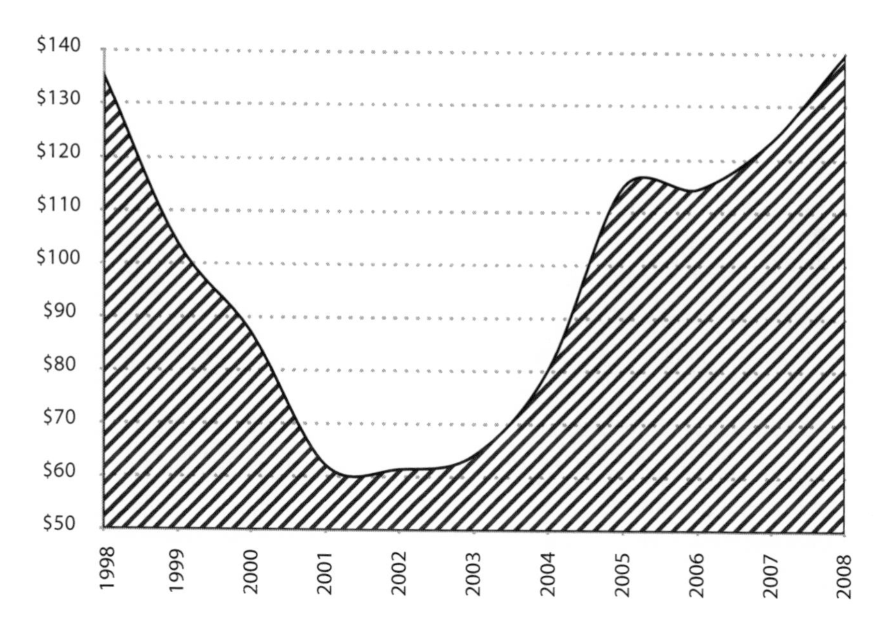

2.1 Yearly Average Coffee Prices (per 100 lbs)
Source: International Coffee Organization

who express this view take for granted that (1) their groups are organized, which most of them were before acquiring Fair Trade certification, (2) they are able to produce export-ready coffee, and (3) they have access to specialty coffee markets. The latter two preconditions were likely achieved via participation in Fair Trade value chains. Recognizing this, the director of one union was more circumspect: "The benefits are not so clear like they were three years ago. However, Fair Trade guarantees us relationships and prices in the event that the market falls. Therefore we continue to work under the system."

Farmer-Members

We surveyed one to four randomly selected farmer-members of nineteen cooperatives. Our questions encompassed their understanding of Fair Trade, the general advantages and disadvantages of being part of a Fair Trade cooperative, and if and how membership had personally benefitted the respondent.

Of thirty-four farmers interviewed, twenty-two (65 percent) exhibited some understanding of Fair Trade based on internationally recognized definitions. Either they could describe it accurately in market terms (e.g., "a fair price for farmers") or they knew what the social premium was and how their cooperative had used it. Three (9 percent) articulated deep knowledge, reflecting both social justice and fair market principles; nine (26 percent) had erroneous or no knowledge. This finding corresponds to what other researchers report; many studies reveal that farmer-members of Fair Trade cooperatives relate to Fair Trade on the basis of price or organic production, not in terms of solidarity, alternative markets, empowerment, or sustainability.

What of the Guatemalan farmers who *did* exhibit a comprehensive knowledge of Fair Trade? All three have responsibilities within their cooperatives; one is a manager, one is in charge of the wet mill (for processing coffee cherries), and one is his group's treasurer. However, other similarly active members of the same cooperatives did not know much about Fair Trade's mission. The exceptionally knowledgeable farmers do not stand out on in terms of their age, years of education, or years of cooperative membership. Rather, it is their answers to the question "Why did you join?" that help explain their multifaceted knowledge of Fair Trade. One man said that he joined for help with converting to organic production and learning about the environment, and because of the Fair Trade social premium. Another mentioned wanting to raise money to support social projects. The third talked about converting his farm from one that "used a lot of chemicals" to one that is certified organic. "Today I am trained in many new things, which keep me motivated as a member." These responses contrast with those of the majority of farmers, many of whom claimed "better price" as their primary reason for joining a Fair Trade cooperative. In these outlier cases, farmers' understanding of Fair Trade relates to their motivation for participating in it.

Table 2.4 summarizes farmers' answers to questions about the advantages and disadvantages of Fair Trade, and their personal experience with it. By far the most frequently mentioned advantage is a stable or better price in a turbulent market. "We do this work because one day the coffee market will drop, and then we will have a certain price." Some farmers note that buyers pay late or that selling via the cooperative is less lucrative when market prices are high (because, assuming they have access to processing facilities, they can get the same amount of money for their crop instantly as opposed to waiting

Table 2.4
Advantages and Disadvantages of Fair Trade as Reported by Participating Farmers (N=34)

General Advantages	**No. of Mentions**
Stable/better price	22
Social premium	6
Working together	4
Health care / better health	4
Pro-environment	4
General Disadvantages	
Late payments	4
FT not competitive when prices are high	2
Too much reliance on family labor	1
Hard to communicate with FLO	1
Personal Benefits*	
Training**	12
Credit	10
Job with the cooperative	6
Education for one's children	6
Extra income***	5
Access to clean water	3
Electricity	2
Learning about prices	1
Increased confidence	1

* All interviewees said they had personally benefitted from participation in Fair Trade.
** Includes training in organic farming.
*** Farmers reported spending extra income on houses (3), farms (2) and training in alternative medicine (1).

for Fair Trade buyers to pay the cooperative). But, criticisms notwithstanding, *every* farmer surveyed said he or she had personally benefitted. The most frequently mentioned advantages were training to enhance quality and the transition to organic production as well as access to credit via the cooperative. "The organics get a much better price, but the work is considerably harder. However, after a while one becomes very aware of the good work that being an organic farmer is." "When I need money, I am able to get loans here." "They provide us with low-interest loans to maintain our coffee."

We asked farmers a hypothetical question: If their cooperatives were to receive an extra premium of two quetzals (about twenty cents) per kilogram of Fair Trade coffee, how should they use the money? Encouraged to name as many ideas as they wanted, five mentioned education for their children, four said "community projects," seven chose cash payments to members, and one suggested a fund to help members with health emergencies. But twenty-six farmers talked about things that would help sustain their cooperatives, such as training farmers in organic methods, paying office staff and harvest workers, improving infrastructure for coffee processing and transporting, and more prefinancing. One suggested using the extra funds to pay for cooperative leaders to attend specialty coffee events abroad.

OBSTACLES REAL AND PERCEIVED

In addition to social, economic, and environmental gains, this study has revealed setbacks and challenges to participation in Fair Trade. But even leaders who were very critical of Fair Trade's shortcomings are hopeful about its future. For example, after complaining about slow payment from buyers, the cost of certification, and poor communication with Fairtrade International and country-specific Fair Trade labelers, a general manager said,

> Fair Trade is a system that undoubtedly supports the farmers. Without them [this cooperative] would not exist. . . . Fair Trade has the challenge to figure out how to manage their prices during times like these when the price is high in the market. Fair Trade has to look at the price of certification, the standards of certification, and the competition that they face in the market. If they don't, they will lose clients. . . . I hope that this information is not used to talk badly about Fair Trade, as this is not how we feel about it.

This respondent wanted to make sure that his comments would be used constructively. He hopes that his critique will inform adjustments to the standards and perhaps lead to increased support to help Fair Trade producers navigate complex international markets.

A farmer-member of another cooperative greatly disliked all the paperwork but said, "The people from FLO try to understand what our situation is, how much things cost. . . . They recommend quality [and] have environmental standards—shade, flora, and fauna standards." Again, this comment acknowledges a positive aspect of Fair Trade as well as "the paperwork problem" evident in other studies.

Side-Selling

> There are a bunch of co-ops in the area where we work, and all of them suffer because of all the people who are laundering money. Plus [coyotes] don't pay IVA [tax], and we do. So they have money to pay a better price. Plus they screw us because they don't look for quality and they will buy anything. You have to be a really dedicated member of the co-op to stay.[9]

Fair Trade buyers pay at least the going market price plus the social premium, but how farmers see this depends on two factors: the amount of the Fair Trade price that actually arrives at individual farms after cooperatives have paid debts and operating expenses (see Jaffe 2007, Lyon 2006), and the relative draw of side-selling. Some members of Guatemalan Fair Trade coffee cooperatives sell a portion of their crops to local intermediaries known as *coyotes*. This usually takes place at the farm level and on a much smaller scale than in the case of the suspended cooperative mentioned earlier. The main reason for side-selling is fast cash, no strings attached. When producers sell to their Fair Trade cooperatives, they often must wait several months to be paid because the cooperative also must wait. In the absence of adequate credit (extended by buyers to cooperatives or by cooperatives to members), therefore, side-selling provides a way to obtain resources needed to pay for food, school fees, and expenses related to coffee production.

Asked how *coyotes* are able to pay instantly, cooperative leaders explain that (especially when the coffee is unprocessed) *coyotes* have access to an economy of scale that has an established infrastructure for processing coffee and a market with buyers ready to purchase, with value added in excess of processing costs. Furthermore, *coyotes* engage in practices that minimize their costs and maximize their cash flow and profits, such as mixing low- and high-quality coffees, cheating on weights, evading taxes, and drug-money laundering.

Side-selling is not unique to Guatemala. A study of Fair Trade and poverty alleviation in Chiapas, Mexico, demonstrated that cooperative members were side-selling out of economic necessity, to cover production costs or their children's school expenses (González Cabañas 2002). Another field study conducted in and around a Fair Trade cooperative in Oaxaca, Mexico, (which included a lunch with the neighborhood *coyote*) found that farmers hold back a certain amount of their export-grade coffee "to use as cash to buy food and other needs from the *coyote* merchants." A cooperative leader told the researcher that this practice "comes from producers' not fully understanding how to maximize their income." However, the leader admitted, "The organization is not right at hand, and the *coyote* is" (Jaffee 2007, 107). Yet another study of the individual Fair Trade cooperatives associated with a northern Nicaraguan second-tier organization reveals that there, large local exporters as well as *coyotes* offer immediate payment to farmers. In 2005, 20 percent of the producers associated with this union breached their contracts, even though they were penalized five cents per pound for doing so (Utting 2009, 13).

Imagine a cooperative that produces only export-grade coffee, is transparently and efficiently managed, has adequate access to credit, and is composed of farmers committed to Fair Trade. Perhaps in this fictitious co-op there would be no side-selling; the Fair Trade value chain would be the *only* value chain the cooperative participates in. But given that side-selling is relatively common in the real world, a most practical question is whether or not side-selling is undermining Fair Trade. We specifically asked farmers about the volume of coffee they sold on the side each year, whether it was unprocessed (cherry) or processed, its quality, and the price they received. Analyzing the data, research assistant Kira Luna (2008) found that—besides the instant cash incentive—farmers sell to *coyotes* because the latter buy coffee of varied quality, including beans that would not be accepted by international Fair Trade buyers. Some cooperatives actually *encourage* members to side-sell their lower-grade coffee. Furthermore, *coyotes* will buy unprocessed coffee and pick it up at the farm, a big advantage to producers who lack reliable transportation from their farms to the cooperative's mill. Luna's investigation also shows that, in each of four harvests, the only coffee for which *coyotes* paid more than the cooperatives was *unprocessed*.

Collectively, farmers reported selling about 33 percent of their coffee to buyers outside their cooperatives.[10] Asked whether they would sell more to *coyotes* if the price increased, 74 percent of the farmers said no. They understood the importance of maintaining stability within the cooperative, knew that they were obligated to sell a contractual amount of coffee to the cooperative, and that the cooperative offered a stable price, unlike the fluctuating street price (Luna 2008, 18). "Sometimes the market is up, and sometimes it is down, but we are getting a stable price." "The *coyotes* can kill us when they see that we have coffee to sell. But via [the farmer's cooperative] we are able to sell it together."

Overall, farmers' actual practices and views suggest that side-selling is not significantly altering farmers-cooperative-buyer relationships in Fair Trade value chains. This is because the coffee sold to *coyotes* is mostly of low quality, and it is a small portion of farmers' total output. Luna concludes that, in the Guatemalan case, "producers are using multiple trade networks to their advantage without jeopardizing the success of their cooperatives" (2008, 20). The best things cooperative leaders can to do reduce or eliminate side-selling are (1) help farmers improve the quality of their coffee, (2) extend low-interest credit, and (3) educate members about the broader meaning of Fair Trade.

Producer Integration

Recall that most of the farmer-members we surveyed did not have much understanding about the Fair Trade system in which they were participating. Our interviews with cooperative leaders indicate that, over the past few years, they have made great efforts to help farmers improve the quality of their coffee—at the behest of buyers and Fair Trade NGOs. This has improved their sales within Fair Trade and other specialty coffee markets, but has generally not imbued farmers with a sense that they are involved in a *movement*. Does this matter? One might reason that it does not, as long as farmers are reaping tangible financial or quality-of-life benefits from Fair Trade. Others (e.g., VanderHoff Boersma 2009) contend that, without producer integration, Fair Trade will ultimately fail as a development strategy and an agent of social change. This research suggests that, in Guatemala, much could be done at the local level to enhance farmers' involvement in Fair Trade.

With prices on the rise, farmers (or cooperatives) that view Fair Trade simply as a link to international specialty markets are likely to sell to the highest, fastest-paying bidder. But coffee markets are extremely volatile. It is entirely possible that prices will again fall to levels so low that it makes more sense to burn coffee as fuel than to harvest it, as was the case for some Central American farmers in the early 2000s. The leaders of one cooperative in the survey, whose coffee has won multiple quality awards, no longer see an advantage to participating in Fair Trade because at present they can sell directly to high-paying buyers without the certification fees and paperwork that Fair Trade entails. But this strategy of using Fair Trade as a temporary measure to facilitate access to specialty markets overlooks the important long-term benefits of Fair Trade: a guaranteed floor price, contractual arrangements with buyers, access to credit, and the social premium.

Managing the Paperwork

Some Guatemalan cooperatives are having a very hard time maintaining efficient communication with Fairtrade International. After describing a long-running misunderstanding regarding inspection fees and reports, one cooperative's executive director remarked, "I don't understand who does what and who I am supposed to talk to there. . . . I think that they are creating a lot of structures, which add costs, even though they *are* trying to create dialogue." This is an issue that Fairtrade International is aware of and seeking to address. Smallholder coffee farmers in Guatemala are generally underprivileged members of their communities, with little if any formal education or access to basic social, educational, or financial programs. They may be expert farmers, but interpreting, implementing, and managing Fair Trade standards is a huge challenge. As expressed by one cooperative leader,

> FLO is created by Germany, and here we have so many people who can hardly read or write and they are expected to justify everything they do via a written form. And they can't put their opinion into what they are being evaluated in. The level of education that they are requiring isn't just to fill out a form with their name, but it is pages and pages of information.

NGOs such as Crecer and Intermom (Oxfam Spain) often subsidize the salaries of general managers, accountants, and consultants. This creates jobs but limits the transfer of professional skills, since the people who fill the jobs usually come from outside of the cooperative and its community. Once their contracts are over these people leave, taking their skills with them. To deal with this problem, Crecer's Fair Trade project has shifted its focus toward training cooperative member-staffers as well as providing qualified human resources (Sanborn 2008).

In 2006 FLO established a producer business unit and a fund to defray the costs of Fair Trade certification. They hired in-country liaison officers in seventeen countries, including Guatemala, and appointed producers to serve on the organization's board. These steps laid the groundwork for better communication and a greater voice for producers within global Fair Trade networks. This study indicates that more work is needed, but also offers reason for optimism. Explaining why his cooperative loyally supplies its Fair Trade customers even when open market prices are high, one leader said,

> We are in this for the philosophy of Fair Trade; it is a different type of friendship and relationship. This is why it is called an "alternative market," so that we can contribute and make it better and improve the world. FLO, buyers, producers, and consumers as well as banks and others are working together in this; this is why we are in this. Every day it is getting better, and we are creating just relationships with one another. [Another buyer] can come and pay us more, but this is just about money, and this will destroy us.

CONCLUSION

Fair Trade contributes to development in Guatemala in material (e.g., funding infrastructure projects) and immaterial ways (e.g., promoting democracy and human rights). Although farmers do not always understand Fair Trade standards, they do express concrete ways in which participation has benefitted them. They also note ways in which the Fair Trade system is problematic and could be refined, either by changing the rules or simply through better communication. For example, one farmer erroneously believed that Fair Trade participants were not allowed to use hired labor.

He complained that he had to take his children out of school to help with the harvest. Another farmer in the same cooperative said, "The people from Fair Trade want the family to be involved in the harvest, however, my children are studying, and so I work when I can in the field, but I need help and need more people to help me. But they don't want us to do this." Fairtrade International's ongoing refinements in response to communication gaffes such as these and "the paperwork problem" do, however, indicate—to me and to the cooperative members who criticize Fair Trade but stick with it—that in Guatemala the Fair Trade coffee cup is half full, not half empty.

What lies ahead? There are opportunities to strengthen Fair Trade from within producer organizations and through their relationships with FLO, country-level Fair Trade NGOs, and buyers. While expanding Fair Trade markets means more proceeds and premiums for producers, growing pains sometimes accompany the push to fill rising demand. A long-time Fair Trade-organic importer of Guatemalan coffee also believes that Fair Trade's shortcomings are a wake-up call: "Fair Trade is going to morph into something else. What's needed is a solution that's lower impact on *all* stakeholders."[11] Given that the primary goal of Fair Trade is to change the norms of global exchange, it would be good news if an end to alternative markets meant that there was no more need for them. In the meantime, learning from Fair Trade's successes and challenges could pave the path from Fair Trade to fair trade.

Chapter 3

How Do Producers Spend the Social Premium?

April Linton and Marie Murphy

An important feature of Fair Trade is the social premium that buyers of Fair Trade certified products pay to the cooperatives or worker organizations that produced the coffee, tea, bananas, rice, wine, and other products they are importing. For example, the Fair Trade coffee premium is twenty cents per pound. Premiums support socioeconomic development as producer-group members see fit. In many ways, it is the social premium that prevents Fair Trade from being "just another market" for producers to sell in.

FLO-Cert, the independent company that certifies Fair Trade products, stipulates:

> The Fairtrade Premium is a tool for development, supporting the organization to realize their development objectives as laid down in its development plan. In the context of small producers' organizations it is meant for investment in the social, economic and environmentally-sustainable development of the organization and its members and through them, their families, workers and the surrounding community. It is for the organization and its members to analyse and evaluate the possible options for spending the Fairtrade Premium. Choices should be made and priorities set depending on the specific situation of the organization and the available amount

> of Fairtrade Premium. Decisions on the use of the Fairtrade Premium are taken democratically by the members, following principles of transparency and participation. It is the joint responsibility of the organization and its members to take wise and fair decisions. (2009, 10)

This statement encourages but does not require Fair Trade producers to allocate premiums to projects that benefit a community beyond their membership. How much do the social and economic benefits of Fair Trade extend from producers to their local communities? What factors influence producer groups' premium spending decisions?[1]

The case studies discussed in chapter 1 tell us a little bit about how Fair Trade producer groups are using the premium, but until now no one has systematically explored this. To do so, we compiled data about producer groups that are actively selling under Fair Trade terms, and then statistically analyzed (1) how and how much they are contributing to projects that benefit the public, and (2) how country- or producer-group specific conditions influence collective decisions to make such contributions. Our cases are Fair Trade plantations, cooperatives, or groups of cooperatives, that is, the entities through which Fair Trade producers trade in global markets. For simplicity's sake, we call all of them producer groups. To collect our data, we systematically combed the websites of Fairtrade International's member groups (table 3.1), looking for information about which producers sell through each labeler, and for "producer profiles." A few producer groups have their own websites; we used information gleaned from them to supplement the profile data, if applicable. Data about five unprofiled Guatemalan coffee cooperatives (used in supplementary analyses only) came directly from our interviews.

We already know that some Fair Trade producers are allocating private resources to public goods in ways that fill gaps left by insufficient state infrastructures and development aid. For example, the Peruvian Cooperativa Huadquiña spends its premiums on environmental initiatives. Oromia Cooperative Coffee Farmers' Union in Ethiopia has built five new schools and upgraded twelve more. They have thirty-six potable water projects and have built eight new clinics with full equipment in areas that had no clinics (Satgar and Williams 2008, 15). The Asociación de parceleros y pequeños productores de bananos (ASOPROBAN) banana cooperative in Colombia

Table 3.1
Fair Trade Labelers Worldwide

Country	Organization	Year Founded
Australia/NewZealand	Fairtrade Labeling Australia & New Zealand	2003
Austria	Fairtrade Austria	1993
Belgium	Max Havelaar Belgium	1991
Canada	TransFair Canada	1994
Denmark	Max Havelaar Denmark	1995
Estonia	Estonian Green Movement—FoE**	2002
Finland	Reilun kaupan edistämisyditys ry	1998
France	Max Havelaar France	1992
Germany	TransFair Germany	1992
Ireland	Fairtrade Mark	1995
Italy	TransFair Italia	1999
Japan	Fairtrade Label Japan	1993
Luxembourg	TransFair-Minka Luxembourg	1992
Mexico	Comercio Justo México A.C.*	1999
Netherlands	Stichting Max Havelaar Netherlands	1988
Norway	Fairtrade Max Havelaar Norway	1997
South Africa	Fairtrade South Africa*	2005
Spain	Asociación para el Sello de Comercio Justo	2005
Sweden	Föreningen för Rättvisemärkt	1996
Switzerland	Max Havelaar Stiftung Switzerland	1992
UK	Fairtrade Foundation	1994
USA	Fair Trade USA (formerly TransFair USA)***	1996

* Associate Members.
** This group makes Fair Trade products available in Estonia but is not a labeler.
*** As of December 31, 2011, Fair Trade USA is no longer a member of FLO.

has started a community garbage-collection and plastic-recycling program and has financed training on environmental protection (Fair Trade USA 2011b). In South Africa, the Wuppertal and Heiveld rooibos tea cooperatives have, in addition to improving their processing facilities, funded local schools (Raynolds and Ngcwanga 2010).

In her study of the Coocafé Fair Trade coffee union in Costa Rica, Lorraine Ronchi remarks:

> Other extensions of Fair Trade benefits can be seen in the services these co-operatives can extend to non-members in the area as a result of improved financial health since Fair Trade involvement. These services range from extending short-term credit at the co-op stores, providing reforestation support and including non-members in housing schemes like the UNIVICOOP scheme run through COOPE CERRO AZUL [a member of Coocafé]. The benefits of Fair Trade, then, appear to be extended to the community at large. (2002, 21)

Further, of the Heiveld rooibos tea cooperative in South Africa, Satgar and Williams note monetary and nonmonetary benefits that stem from participation in Fair Trade. The cooperative has also sponsored community projects, provided an educational scholarship, and provided funds to the local school and church. In addition to the monetary forms of success, the cooperative has helped build a sense of community among farmers scattered across wide distances in a very rugged and disconnected region of South Africa. The sense of solidarity and commitment to uplifting an entire community is perhaps one of the most significant achievements of the Heiveld Co-operative. It has successfully linked alternative forms of production at the local level with a global movement for Fair Trade practices and in the process has begun to build community based on solidarity (Satgar and Williams 2008, 79).

FAIR TRADE AND PUBLIC GOODS

Economists have long proposed that when states fail to provide essential or desired public commodities, local communities make up for the failings in state provisions. We understand public commodities, or public goods, to be socially determined; some goods (e.g., public parks, basic education, and clean water) are made available to all because they are good for society. Many people need these things, but they have differing abilities to pay (Kaul et al. 2003).

The social premium derived from Fair Trade could finance local public goods that are not efficiently provided by states.[2] But while Fair Trade producer groups may use their privately earned premiums to finance local public goods, it is also possible for them to create "club goods" just for themselves (see, e.g., Buchanan 1965, Cornes and Sandler 1996). For example, a group might decide to invest directly in the health or education of its members rather than fund a clinic or school that is available to the entire community, bettering some but potentially creating new inequalities. As Fair Trade expands its reach, it is important to ask, Is Fair Trade creating "islands of wealth" in poor communities? (Kruger and Du Toit 2007, 214).

Why would Fair Trade producer groups distribute their premiums in ways that benefit people who are not part of the group? Recent experimental research shows that members of "clubs" willingly tolerate overcrowding (caused by their voluntarily admitting new members in excess of the club's optimum size) because *they do not want to exclude people they know* (Crossen, Orbell, and Arrow 2004), and that decision makers are most likely to cooperate in a public goods dilemma when they perceive themselves as being similar to their interaction partners (Parks, Sanna, and Berel 2001). In rural farming communities it is likely that people know each other and share similar circumstances, regardless of whether or not they are members of a Fair Trade group.

Other experimental research indicates that people are most likely to contribute to public goods *when they believe that the distribution of goods is fair* (Van Dijk and Wilke 1995; Wit, Wilke, and Oppewal 1992). Participants defined "fair" as a good's being distributed either equally or on the basis of need, regardless of ability to pay (Biel, Eek, and Gärling 1997; Eek, Biel, and Gärling 1998; Eek and Biel 2003). Members of the Nicaraguan Fair Trade cooperatives that Pirotte and his colleagues (2006) studied expressed great confidence in their cooperatives' trustworthiness and in the fairness of their relationships with international buyers. This finding is not necessarily representative of all Fair Trade producers, but when and where producers do believe that Fair Trade is really fair, and to the extent that a sense of fairness can be linked to willingness to contribute to public goods, it seems unlikely that members of a democratic group of Fair Trade producers would vote to spend all their premiums on club goods when an alternative strategy could serve the cooperative *and* the larger community.

Still other studies suggest that contributions to the public's well-being are more forthcoming from individuals who feel accountable to others because they are subject to public scrutiny and want to maintain a good reputation (Gächter and Fehr 1999, Milinski, Semmann, and Krambeck 2002). Building on this idea, David De Cremer's (2003) experiments identify *feeling respected* as the noneconomic motive leading to intentions such as having a good reputation and having good relations with others. In addition to receiving economic rewards, participant observers working in Fair Trade cooperatives similarly connect the empowerment that stems from selling a socially conscious product at a fair price with the respect imparted by these transactions. For example, in their study of a union of Mexican coffee cooperatives, Aranda and Morales note that "the fact that CEPCO sells in the [Fair Trade] markets gives a certain prestige since it is assumed that the organization is subject to external monitoring and also demonstrates initiative and a capacity to enter new market niches" (2002, 16). Regarding another Mexican union, Garza and Trejo observe an increase in self-esteem among the members, "manifested in an increased desire and interest in continuing as farmers who provide the food for their families and also produce coffee commercially. . . . A number of producers have also expressed the pride they feel in belonging to an organization that, in spite of the current price problems, has been able to continue improving its infrastructure and providing assistance through development projects" (2002, 19).

The research discussed above suggests that Fair Trade producer groups will be more likely than not to spend some of their premiums on benefits that extend beyond their groups. And, indeed, our data reveal that a majority (62 percent) of them contribute to projects with more broadly accessible benefits. But how they allocate the social premium may vary in line with their country's level of development, because this influences the perceived and actual need for private contributions toward public services. Public-oriented spending could also reflect characteristics of particular producer groups. Those with longer histories of involvement in Fair Trade may garner more respect in their communities; their members may also have a better understanding of the Fair Trade movement within a social justice and empowerment framework—factors that could positively influence producers' decisions to contribute to public goods.

Since, in developing countries, public goods are often financed with aid from abroad, we are interested in how aid plays into this story. More aid might mean less need for private contributions to public goods, but there is evidence of NGOs (often funded by USAID or the European Commission) and Fair Trade producer groups *partnering* to achieve community development goals. In Nicaragua, "households reported that the [Fair Trade] cooperatives helped them link to NGO-led and importer/roaster sponsored community development projects, including scholarships for education, coffee quality training, and micro credit programs" (Bacon et al. 2008, 267). The environmental NGO SalvaNatura has partnered with Salvadoran coffee cooperative union APECAFÉ to help farmers transition to organic production.[3] González Cabañas's study of the La Selva coffee cooperative union in Mexico reveals ties, at one time or another, to eight different NGOs (2002, 13). In Nicaragua, Fair Trade is part of a larger development project, with lots of NGO participation (Pirotte, Pleyers, and Poncelet 2006).

We note the difference between individual cooperatives or plantations and unions of cooperatives. The governing bodies of the latter sometimes find it difficult to reach a decision about how to spend the premium, which could lead them to follow the least-controversial pathway: spending on group-specific projects, such as internal infrastructure or credit programs. But in some unions, member cooperatives make independent decisions about how to spend their premiums (Aranda and Morales 2002, Ronchi 2002). To the extent that this situation prevails, there should be no difference between spending decisions made by union members and those made by individual cooperatives or plantations.

Finally, we account for producer groups' size (annual production) and success at selling their produce via Fair Trade channels. Larger or more successful producer groups might take on more public projects because they have more premiums to work with.

DATA AND ANALYSIS

The 221 Fair Trade producer groups for which we were able to obtain data are not a random sample. Rather, they are a "population" of groups that have been established long enough to have representation in at least one market served by a Fair Trade NGO and to have earned a social premium. Fair Trade labelers' websites provide information about which producers sell through each labeler, and producer profiles, which include information about producers' locations, years of establishment and Fair Trade certification, annual production, and how members choose to spend the premium.[4] According to Fair Trade USA's Paul Rice and Meghan Quinlan,[5] producer profiles are, to date, the best source of information on producer characteristics and premium allocation. Profiles may be somewhat biased in their emphasis on the positive, but they are not biased toward public goods spending, nor do they contain untrue reports thereof. This is the case because the consumers and activists who read producer profiles are not engaged in arguments about the relative merits of public-oriented or club-oriented development outcomes; they are just as happy to support a system that helps farmers in the global South improve their lives and send their children to school as one in which producer groups invest premiums in community development projects.

How Fair Trade Producers Spend the Social Premium

At some point, all producer groups reinvest the Fair Trade premium into their businesses. They buy trucks and processing equipment, educate members about organic farming and best practices, build storage facilities, and so on. Where else does the money go? Thirty-eight percent of the groups spend it all on themselves. They pay their children's school expenses, provide members-only health services, improve their homes, initiate and maintain credit programs, and sometimes simply divide up the cash. But the rest spend something on public goods. Three categories of investment stand out: education, such as hiring more teachers or building a school (53 percent); providing or improving infrastructure, such as improving roads and water systems or bringing in electricity (46 percent); and health, such as building a community clinic (31 percent). For example, the ASASAPNE cooperative in Guatemala, which sells Fair Trade coffee and honey, uses premiums to repair local schools. In Peru the APARM coffee cooperative has built local roads, the CAC Pangoa cooperative has used premiums from Fair Trade coffee and cocoa to bring electricity to their locale, and the CECANOR coffee cooperative has financed highway repairs. Herkulu Tea Estate in Tanzania has built a new dispensary in the village of Kewhengala, with bimonthly visits from the district medical officer and a health outreach program. There is clearly a preference for spending some of the Fair Trade social premium on public goods, but what factors contribute to this preference?

Contexts for Decision Making: Who Spends on What?

Table 3.2 summarizes the measures we describe in this section. There are competing measures of "development," and none of them is perfect. We chose the Human Development Index (HDI) because it is fairly comprehensive and because the U.N. has generated a HDI score for almost every country on the planet.[6] The HDI includes indicators of three basic dimensions of human development: a long and healthy life (life expectancy), knowledge (education), and a decent standard of living (income) (UNDP 2011). The producer groups in our data are in countries where the 2005 HDI was as low as 0.406 (Ethiopia) to as high as 0.867 (Chile; UNDP 2008). We use the World Bank's 2006 figures to measure international development aid as a percentage of a country's gross domestic product. On the low end are Cuba and Thailand (0 percent); countries that received the most aid (17 to 53 percent) are Ethiopia, Democratic Republic of the Congo, Rwanda, Malawi, and East Timor.

We estimate producer groups' success at generating premiums in two ways. First, many groups sell in multiple Fair Trade markets and are thus profiled by more than one of the labeling NGOs listed in table 3.1. To estimate how much a group actually sells under Fair Trade terms we counted how many labelers list that group as a supplier of one or more Fair Trade products sold in their country. For instance, the Colombian banana cooperative ASOPROBAN sells Fair Trade bananas in Canada, Germany, the Netherlands, the United States, and Switzerland. Second, we look at groups' annual production because more produce, if sold under Fair Trade terms, means more premiums.[7]

As a potential measure of community status and knowledge of/commitment to Fair Trade's goals, we note the number of years that a producer group has had Fair Trade certification. We also note whether or not a producer group is a union. After examining how each of these factors relates to spending on public goods, we explore ways in which various types of spending may be interrelated (because producer groups often support more than one public project).

Table 3.2
Descriptive Statistics, Fair Trade Producers (N=221)

	Minimum	Maximum	Mean	Std. Deviation
Use of Fair Trade Premiums				
Club goods only (N=84)	0	1	0.38	-
education for members' children	0	1	0.48	-
health care for members	0	1	0.36	-
women's initiatives for members	0	1	0.19	-
other services for members	0	1	0.84	-
Public goods (N=137)	0	1	0.62	-
education (general)	0	1	0.53	-
infrastructure	0	1	0.46	-
health (general)	0	1	0.31	-
women's initiatives (general)	0	1	0.17	-
Country-Specific Variables				
Human Development Index (HDI), 2005	0.41	0.87	0.72	0.11
Official development assisance & aid, 2005 (% of GDP)	0.00	52.90	3.78	6.36
Producer Group–Specific Variables				
Fair Trade markets	1	13	3.94	2.58
Annual production (3 categories)	1	3	1.90	0.78
Years of Fair Trade certification	<1	21	8.00	4.96
Union of cooperatives	0	1	0.27	-

Note: For dichotomous variables, 1 = yes.

Findings: What Influences Spending on Public Goods?

The tables in this section detail the results of a logistic regression analysis of spending premiums on public goods. Significant odds ratios of less than zero signify negative relationships; those above zero mean positive relationships. Table 3.3 shows that Fair Trade producer groups are more likely to contribute to public goods in relatively less-developed countries. In places where the HDI is higher, groups tend to keep the premium for themselves, probably because there is not a compelling need for private-sector participation in broader social provision. Development aid and public goods spending are *positively* related. As noted above, many Fair Trade producer groups receive aid, and some have partnered with government development agencies or NGOs to achieve specific projects. In her study of the Nicaraguan Fair Trade coffee union Soppexcca, Karla Utting observes that the group plays a leverage role in helping its cooperatives to "acquire from international agencies or buyer companies the funding they need as a way to complement the fair trade premium and to achieve greater change within fair trade" (2009, 138). In effect, the premium is seed money to get a development project started and then attract donors.

Older producer groups are more likely than their younger counterparts to spend premiums on projects that benefit a wider community. There may be a relationship between years of Fair Trade experience and a sense of respect (which in turn should increase the probability of taking on public projects). It is important to also acknowledge that newer entrants to Fair Trade markets generally have not accumulated enough premium funds to contribute to a public project. Figure 3.1 shows the relative magnitude of the relationships discussed above.[8] There is also a small but significant interaction in the way that development aid and years of Fair Trade experience influence public goods spending. As the level of development assistance rises, the importance of the relationship between experience with Fair Trade and spending on public goods is slightly reduced.

Table 3.4 and the corresponding bar graphs show how the factors discussed in this chapter's previous section relate to *specific types* of public goods spending—on education, public health initiatives, and local infrastructure. In Figure 3.2, the strongest predictor of Fair Trade producer groups using premiums on general education (building schools, hiring teachers, buying classroom supplies, and so on) is the country-level HDI; where governments are least

Table 3.3
Odds Ratios for Allocation of Fair Trade Premiums on Public Goods (N=221)

	I	II	III	IV
Country-Specific Variables				
Human Development Index (HDI)	0.109	0.045#	0.011#	0.040#
development aid	1.088*	1.084#	1.081#	1.209**
Producer Group–Specific Variables				
FT markets		1.060		
annual production		1.144		
years of FT certification		1.099**	0.959	1.163***
union of cooperatives		0.843		
Interactions				
years of FT x HDI			1.231	
years of FT x development aid				0.986*
Constant	6.520	3.940	15.019	4.294
Chi-square	13.542***	27.982***	27.136***	30.418***
-2 Log Likelihood	227.960	263.520	264.366	261.084
Percent correct	63.3	65.6	67.0	65.2

# p <.10 * p <.05 ** p <.01 *** p <.001

likely to adequately fund education, Fair Trade groups are most likely to do so. The only producer group–level characteristic related to spending on general education is the amount of time the group has been Fair Trade certified. Interestingly, producer groups that invest in general education are also likely to be involved in public health and infrastructure projects.

Figure 3.3 shows that producer groups are most likely to spend premiums on public health (building clinics, hiring staff, providing health education,

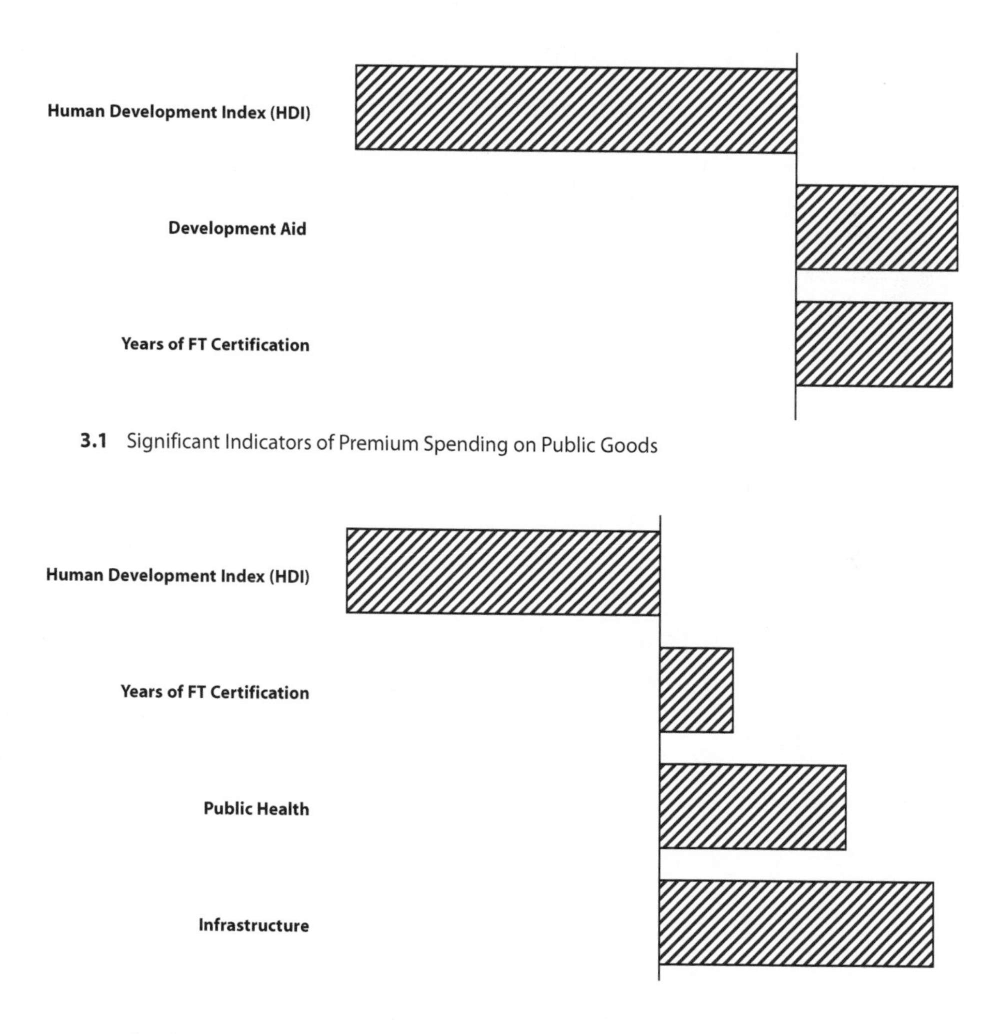

3.1 Significant Indicators of Premium Spending on Public Goods

3.2 Significant Indicators of Premium Spending on General Education3

and so on) in countries with relatively high development aid—again suggesting cooperation between aid agencies and Fair Trade producer groups. Groups that invest in public health tend to be relatively large, and independent organizations (rather than unions). They are likely to also be spending on general education, and *not* spending on members-only health initiatives (e.g., health insurance or exclusive clinics).[9]

Figure 3.4 documents that, not surprisingly, Fair Trade groups in rela-

Table 3.4
Odds Ratios for Allocation of Fair Trade Premiums on Public Goods (N=221)

	General Education		
	I	II	III
Country-Specific Variables			
Human Development Index	0.007**	0.002***	0.011**
development aid	0.987	0.977	0.979
Producer Group–Specific Variables			
FT markets		1.062	1.093
annual production		1.157	1.070
years of FT certification		1.100**	1.072#
union of cooperatives		0.911	0.947
Other Public Goods Spending			
general education			
public health			2.670*
infrastructure			4.289***
Club Goods Spending			
members' health care			
other services for members			
Constant	17.185**	11.311*	2.217
Chi-square	13.282***	28.378***	53.156***
-2 Log Likelihood	267.124	252.028	227.250
Percent correct	70.6	69.7	73.3

# p <.10 * p <.05 ** p <.01 *** p <.001

Public Health				Infrastructure			
I	II	III	IV	I	II	III	IV
0.152	0.109	0.478	0.494	0.007***	0.003***	0.013*	0.012**
1.052#	1.059#	1.071*	1.065*	0.969	0.963	0.966	0.966
	1.075	1.066	1.086		0.900	0.864#	0.856#
	1.562#	1.518#	1.548#		1.018	0.971	0.928
	1.077#	1.056	1.046		1.094*	1.073#	1.083*
	0.305*	0.297*	0.266*		1.348	1.453	1.548
		2.630*	2.647*			4.317***	4.299***
						1.090	1.131
		1.202	1.217				
			0.417#				
							0.523#
0.744	0.197	0.055#	0.068#	15.507*	14.040*	5.203	8.800
9.722**	23.578***	30.752***	34.867***	11.201**	18.698**	37.740***	40.749***
210.900	197.045	189.870	185.755	254.788	247.292	228.250	225.241
81.0	81.4	81.9	83.3	73.3	72.4	76.5	74.7

# p <.10 * p <.05 ** p <.01 *** p <.001

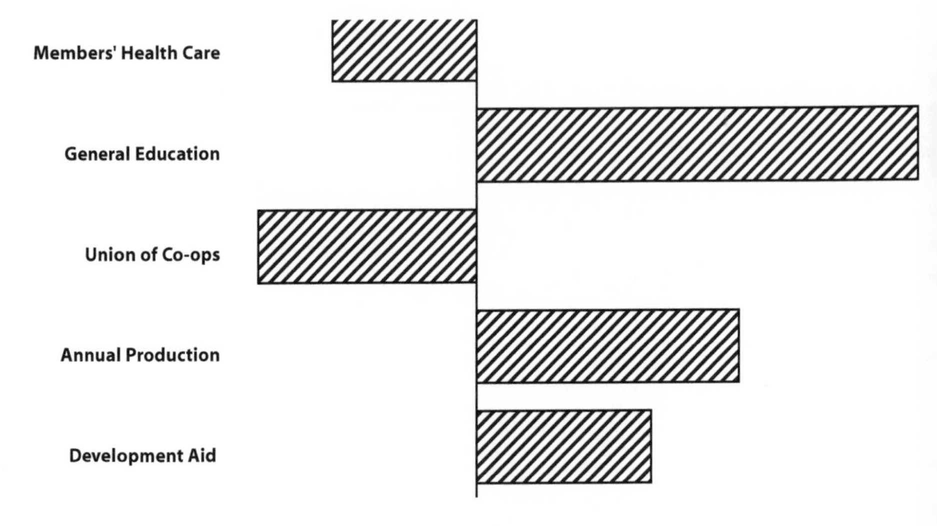

3.3 Significant Indicators of Premium Spending on Public Health

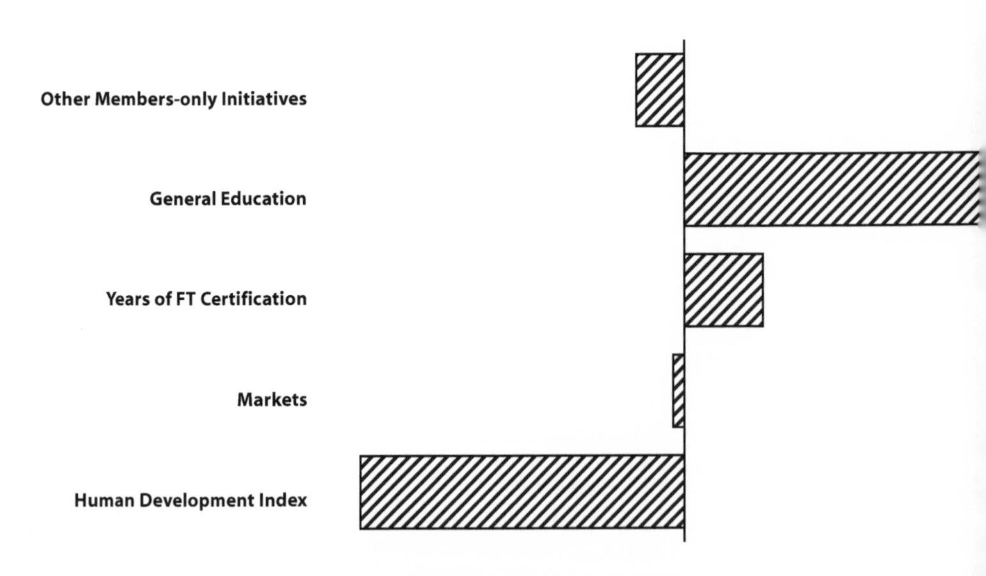

3.4 Significant Indicators of Premium Spending on Public Infrastructur4

tively less-developed countries are more likely to spend their premiums on community infrastructure projects such as delivering electricity or improving roads and access to clean water. Groups that spend premiums on public

Table 3.5
"Other Services for Members" (N=21)

	2nd-tier	Primary	Total	Percent
Credit	2	4	6	29
Home improvements	2	2	4	19
Pay loans or for coffee	2	2	4	19
Cash disbursements	4	3	7	33
Life insurance	0	1	1	5

Note: Only studies that mention something coded as "other services" are included.

education are more likely to spend on public infrastructure as well, but spending on infrastructure is about half as likely when a group is dedicating premiums to "other" members-only endeavors. To understand what it means to spend on "other services for members" (which, in producer profiles, is almost always represented as "home improvements" or "credit programs"), we revisited the field studies (from chapter 1) and the Guatemala surveys (from chapter 2).

In Table 3.5, better housing and credit do stand out as components of "other" members-only spending, but a third of the case studies and interviews mention direct cash payments. Some groups disburse the premium in this way because the amount seems too small for a collective project. For example, the administrative manager of one Nicaraguan union said, "It is difficult for consumers as well as producers to see where the fair trade premium goes because the market of fair trade is relatively small here. It is not easy to promote big changes when we only export 10,000 quintales of fair trade coffee, which gives us approximately US$50,000 per year" (quoted in Utting-Chamorro 2005, 595).[10]

Other producer groups divide the premium because this is what members want. The leader of a Guatemalan cooperative that has previously spent premiums on education, health care, farm improvements, and buying more property reported members' voting to change these practices: "They all agreed on receiving their social premium as individuals instead of a group. The people want to see the cash." This is—at least in part—because of inadequate prefinancing from buyers. Fairtrade International does not support this practice, but its existence has, in recent years, brought the issue of prefinancing to the fore as an aspect of Fair Trade in need of greater enforcement. As of February 2009, the revised Fair Trade criteria include a statement that the use of the premium must be democratically decided *in advance* and the premium must be spent according to this work plan.

WHAT CAN WE LEARN FROM THIS ANALYSIS?

Fair Trade spurs private contributions to public well-being. The majority of Fair Trade producer groups are contributing to the well-being of their communities and regions by building or supporting schools, wells, roads, and health care facilities. Younger producer groups and those in more-developed countries are less likely to spend any of their Fair Trade premiums on public goods, whereas groups in countries receiving large amounts of development aid are more likely to do so. Fair Trade producer groups that allocate funds towards general education tend to be in countries with relatively low HDI scores, have more experience with Fair Trade, and also spend on community health and infrastructure. In table 3.6, we see a comparison between the entire data set and groups that make public contributions to (1) education, health, and infrastructure; (2) education and health; (3) education and infrastructure; and (4) health and infrastructure. This table shows that producer groups that spend premiums on multiple public-goods projects are unique in some ways. Groups that are heavily invested in providing public goods are, relative to the entire population of active Fair Trade producers, located in very poor countries, and (for spending on health, education, *and* infrastructure) in countries where a fairly high percentage of the GDP comes from international aid. For example, the average HDI score across producer groups is 0.72, but for groups that are working on education, health, *and* infrastructure projects it is only 0.56. The overall average years of Fair Trade certification is 8.04, but groups that invest in general education *and* public health average 11.92 of Fair Trade experience.

Overall, these findings indicate that Fair Trade producer groups are establishing and prioritizing local development goals, especially in the least-developed countries, and that older Fair Trade groups are more likely to

Table 3.6
Mean Independent Variable Levels for Contributors to Public Education, Health, and Infrastruture

Public Goods Contributions	**All Three**	**Ed. & Health**	**Ed. & Infra.**	**Health & Infra.**	**All Cases**
Independent Variables	N=12	N=13	N=26	N=5	N=221
HDI	0.56**	0.72	0.67#	0.79***	0.72
Development assistance	10.63*	3.22	2.51	2.04	3.78
Annual production	2.17	2.23	1.73	2.40	1.90
Years of FT certification	9.42	11.9**	8.50	11.40	8.00
Fair Trade markets	4.75	5.07	3.88	4.00	3.94
Union of cooperatives	0.25	0.15	0.27	0.20	0.27

#p <.10 * p <.05 ** p <.01 *** p <.001
Note: Significant differences are noted in comparison to the reference category "all cases."

contribute to the public good than their younger counterparts. There is a positive relationship between development aid and public goods spending; it appears that Fair Trade *and* aid is a winning combination for farming communities in the global South.

Most of the time, size was not an important predictor of a Fair Trade producer group's contributions to public goods. One exception is that larger and independent groups (as opposed to unions) are more likely to allocate premiums to public health initiatives. This may be because building and staffing a clinic or mobile health facility is an expensive endeavor that only larger producers can afford. A centralized decision of this magnitude could be hard to reach agreement on, and where individual member cooperatives in unions make independent decisions, they are not likely to have sufficient resources.

Market access and annual production do not matter in terms of the general decision to spend premiums on public goods, and annual production is significantly related only to spending on public health initiatives. We initially reasoned that groups with greater Fair Trade penetration would be likely to sell more of their produce under Fair Trade terms and thus generate more premiums proportional to their annual production. Also, working with several labelers would increase groups' interaction with the Fair Trade system, which encourages them to pursue a development agenda. However, it appears that our measure of "markets"—how many countries a group sells Fair Trade produce in—does not adequately capture their success because the markets are of unequal size and because it is not possible to know how much groups are actually able to sell under Fair Trade terms.

Two additional insights about the importance of quality and the less-tangible benefits of Fair Trade come from the Guatemala study in chapter 2. Of the twenty groups surveyed, only three reported current spending on public goods (repairs and new computers for schools, teacher training, and a health center). These three groups stand out in a way we were not able to measure here: their coffee is of exceptionally good quality. All three produce organically. Their leaders actively work with members to improve quality, and the cooperatives have access to cupping facilities. At the time of the survey, these three were receiving prices higher that what Fair Trade stipulates; one group reported getting $1.92 per pound. The other seventeen groups chose to use the social premium on things, such as better coffee-processing facilities, scholarships for their children, microloans, health insurance, or home improvements for members. Some farmer-members use their shares of the premium simply to keep their families financially afloat.

There are limitations to the conclusions we can draw from this research. First of all, there is considerable geographic variation in the overlap of "producer group" and "community." In some cases (especially unions) there is little or no overlap; in others, such as tea estates, the Fair Trade producers *are* the community. Our own observations and the case studies we reviewed suggest that most situations fall somewhere in between these extremes. Another limitation is that we do not have data about how much money went to the reported projects. Further research along these lines would benefit from more in-depth information about premium spending. We would also like to see local (rather than country-specific) indicators of development and aid and a better measure of how much a group is selling to Fair Trade buyers.

CONCLUSION

This study has shown that, in addition to benefitting farmers and their families, Fair Trade is benefitting entire communities in the global South. In a time when (to the consternation of some Fair Trade activists) Fair Trade's success is largely measured in terms of sales, we show that producer groups successfully selling via Fair Trade channels are actively promoting community development through their use of the Fair Trade social premium. We should also note that Fair Trade's contribution to development is more than economic. In their study of Fair Trade coffee cooperatives in Nicaragua and Tanzania, Gautier Pirotte and his colleagues point out that the cooperatives nurture a sense of solidarity just as they train producers and advocate an entrepreneurial spirit: "Fair Trade has given small-scale producers the economic security to enable them to develop and take charge of their own lives within the cooperative network" (Pirotte, Pleyers, and Poncelet 2006, 450). In other words, relative economic security is contributing to Fair Trade producers' solidarity. After documenting many successful projects undertaken by the Mexican cooperative union UCIRI, Francisco VanderHoff Boersma remarks on the link between "increased capacity building in the economic and social realms" and "empowerment in the political realm" (2009, 56); for example, members of the cooperative union have pressured the government for infrastructure improvements. Wisely managed, Fair Trade could continue to play an important role in creating and expanding international market relationships that incorporate sustainable development and social responsibility.

Chapter 4

Selling and Buying Fair Trade

"Fair Traders"—labelers,[1] activists, committed businesses, and other organizations—promote Fair Trade Certified products, but what they are really selling is a social value: putting your money behind Fair Trade will help alleviate severe social inequalities faced by farmers in less-developed countries. Using a label to achieve this end works because consumers widely sympathize with the ethical framework the label represents (cf. Seidman 2007). But in his study of anti-sweatshop and sustainable forest products initiatives, sociologist Tim Bartley reminds us that certification and labeling initiatives did not emerge as *responses* to the rise of socially and environmentally responsible consumerism; stable markets for certified products rarely exist before the certification programs are initiated. Creating new markets is "part of a larger institution-building project that occurs along with the construction of certification associations" (2003, 435). This has been the case with Fair Trade. It was necessary to popularize the concept and the products by channeling existing consumer demand for ethical choices, and by working to increase that demand.

A wide array of social science, consumer studies, and marketing literature helps us evaluate ways in which Fair Traders help businesses and end consumers learn about what Fair Trade is and encourage them to make a connection connect between sympathy for a cause and actual behavior. This

4.1 Past and Present Fair Trade Labels

chapter summarizes important pieces of that literature, concluding with a discussion of Fair Trade as political consumerism.

TARGETING BUSINESSES

Customers, shareholders, financial markets, and even the insurance industry are asking more of companies than they used to in terms of social and environmental accountability (Grodnik and Conroy 2007). A recent survey by the global accounting firm KPMG (2009) showed that, in 2008, 80 percent of Global 250 companies issued a public report on corporate responsibility. Options for socially responsible consumption are ever increasing, and the Internet has become a tool for rapidly spreading information about companies' ethical practices, including charges of social or environmental irresponsibility.

Fair Traders' business-focused campaigns began by targeting companies that were at the forefront of socially responsible commerce. In Europe, these included large roasters, such as Neuteboom and La Semeuse. In the United States, roasters such as Equal Exchange, specialty stores emphasizing "natural" and environmentally friendly products, and independent coffee shops were among the first to start purchasing Fair Trade coffee. Other businesses followed suit; in 2000, TransFair USA (now Fair Trade USA) entered into a contract with Starbucks, the world's largest coffee chain.[2]

Fairtrade International and the Fair Trade labelers do not require businesses that sell Fair Trade products to market them in any particular way. Some businesses aggressively promote Fair Trade, others source Fair Trade products but do not broadcast the fact, and still others offer Fair Trade options as if they were flavors. There is a role for Fair Trade in "cause-related marketing"—companies' efforts to link themselves to relevant social issues. Consumers are known to attach emotionally as well as practically to companies and brands (Pringle and Thompson 1999); businesses can therefore promote brand attachment by using Fair Trade to add to their brand's "soul." This strategy, however, carries a challenge: Fair Trade is already a separate "brand" to which competitors have access. To address this, companies must align their Fair Trade marketing focus with an experience that customers want to have. For businesses that already enjoy strong brand loyalty from their customers, the Fair Trade label could become a "brand extension" or a part thereof (David 2000, 132; Linton, Liou, and Shaw 2004).

In the United States, Whole Foods Market's "Whole Trade" project is an example of a brand extension. According to the company's website, Whole Trade is "a commitment to ethical trade, the environment, and quality products" (Whole Foods Market 2009a). The company currently works with Fair Trade USA, the environmental NGO Rainforest Alliance, and the Institute for Marketecology ("Fair for Life" label).[3] This multilabel partnership enables Whole Foods to brand a broad array of goods, including textiles and body care products, with the Whole Trade seal. But of the three labels, only "Fair Trade" guarantees producers a predetermined floor price and social premium.

> We work with a variety of certifiers to accommodate the breadth of the Whole Trade Guarantee initiative and to foster healthy competition while continuing to expand the variety of products involved. Our partners certify products to meet the four key criteria required by Whole Trade Guarantee: quality, premium price to the producer, better wages and working conditions, and the environment. (Whole Foods Market 2009b)

Whole Trade is part of Whole Foods' ongoing effort to responsibly source everything the company sells, be it locally grown organic produce, meat from humanely raised animals, or cleaning products made from nontoxic, biodegradable materials. When sourcing imports from the developing world, the company does not single out Fair Trade as the only socially responsible label but rather allows customers to identify a broad array of branded products that have been produced and obtained through socially responsible means. As Whole Foods' grocery support and Whole Trade specialist Jessica Hasslocher Johnson has emphasized, the company's interest is in branding its own program, not simply promoting the certifiers they work with. In order to maintain "healthy competition" among certifiers, give more growers access to the Whole Trade program, and maximize customers' product options, Whole Foods intends to expand the list of Whole Trade certifiers year after year.[4]

Whole Foods is following a tactic similar to Starbucks' Commitment to Origins, which emphasizes growers' receiving a good price for high-quality coffee as well as social investment in coffee-producing countries. Both programs aim to provide an overall shopping experience in which many sustainably sourced products are on offer. Some Fair Traders do not support this approach because it makes retailers' sourcing look more ethical than it really is, and—at least potentially—pits various ethical labeling initiatives against each other. Companies' response to this criticism is that they are trying to progressively improve the sustainability of their supply chains while meeting their customers' demands. Given that, as Johnson put it, "the American consumer is still pretty uneducated about trade," this argument is reasonable. When relatively large companies such as Whole Foods and Starbucks source what is for them a small quantity of products under Fair Trade terms, it can make a big difference for producers. Fair Trade purists still have options in regard to voting with their money; they can patronize firms such as Whole Foods and Starbucks but buy only Fair Trade Certified coffee, tea, chocolate, and such, or opt to support smaller, 100 percent Fair Trade retailers.

Another point of contention is that Fair Trade USA requires less of a commitment to Fair Trade as a proportion of total sales volume (of coffee, for example) from large retailers than it does from smaller companies. Fair Trade USA justifies this decision by reasoning that (1) a smaller percentage is still a very large volume, which means more money for the producers, and (2) the big companies need to "test the water" to assess their customers' demand for Fair Trade. If the demand is there, Fair Trade purchases will increase (Jaffee 2007, 203–4).

RESEARCH ABOUT BUSINESSES AND FAIR TRADE

The studies summarized below critically examine companies' motivations for promoting Fair Trade, management and marketing strategies, grocery-store brand Fair Trade products, and Fair Trade business networks. The businesses studied are Green Mountain Coffee Roasters in the United States; Coop, Switcher, McDonald's, and Magasin du Monde in Switzerland; and Asda, Co-op, Morrisons, Safeway, Sainsbury's, Somerfield, Tesco, Waitrose, Equal Exchange, Cafédirect, and the Day Chocolate Company in the United Kingdom.

Why Join the Movement?

Environmental finance expert Ann Grodnik and economist Michael Conroy tell the story of how one coffee roaster came to enthusiastically "focus on dual-certified Fair Trade and organic coffees as the most rapidly growing component of its total sales" (2007, 83).[5] Founded in 1981, Vermont-based Green Mountain Coffee Roasters (GMCR) roasts premium-quality coffees and has always been known for its long-term relationships and fair dealings with suppliers in Central and South America and Indonesia. Before Fair Trade certification was available in the United States, Green Mountain developed its own line of socially and environmentally friendly coffee and launched some organic varieties, but these were sidelines, mainly because of the challenge of implementing such a program and the fact that most Green Mountain coffee was being sold in mainstream supermarkets—not a particularly good outlet for a "responsible label."[6]

Green Mountain began to significantly integrate and strategically market its social and environmental priorities in 2000, when the company acquired an organic brand and was subsequently introduced to Fair Trade. As a promotion, they set up a motor home as a coffee shop on wheels. The "Buzz Mobile" traveled around New England, giving away Fair Trade coffee and educating patrons. Buzz Mobile staff took every opportunity to talk to the press and to speak at conferences or at colleges. Meanwhile, Green Mountain's leaders "decided that this had to be a win-win to build volume with producers and for GMCR"—a way to attract new customers.[7] They decided to price Fair Trade Certified coffees in the middle of their line so that it would be viable to put the coffees in supermarkets' big bin coffee displays. This turned out to be a great way for stores to differentiate themselves from their competitors. Meanwhile, GMCR trained its sales force about Fair Trade and about the company's commitment to it. These salespeople in turn worked with the international social justice and antipoverty NGO Oxfam to educate university food service managers.

Three years after GMCR introduced them, Fair Trade coffees comprised 13 percent of the company's offerings (Grodnik and Conroy 2007, 92). In 2008 the figure was 28 percent.[8] Green Mountain now aggressively markets Fair Trade, often featuring detailed "notes from origin" about coffee producers and their communities. How did the company justify this move to its shareholders? First of all, the company's leaders argued that entering Fair Trade value chains is an investment in sustained supply and a way to standardize and monitor quality. Second, this strategy could differentiate Green Mountain from its competitors and extend the company's New England market reach to the national level via partnerships with nationwide

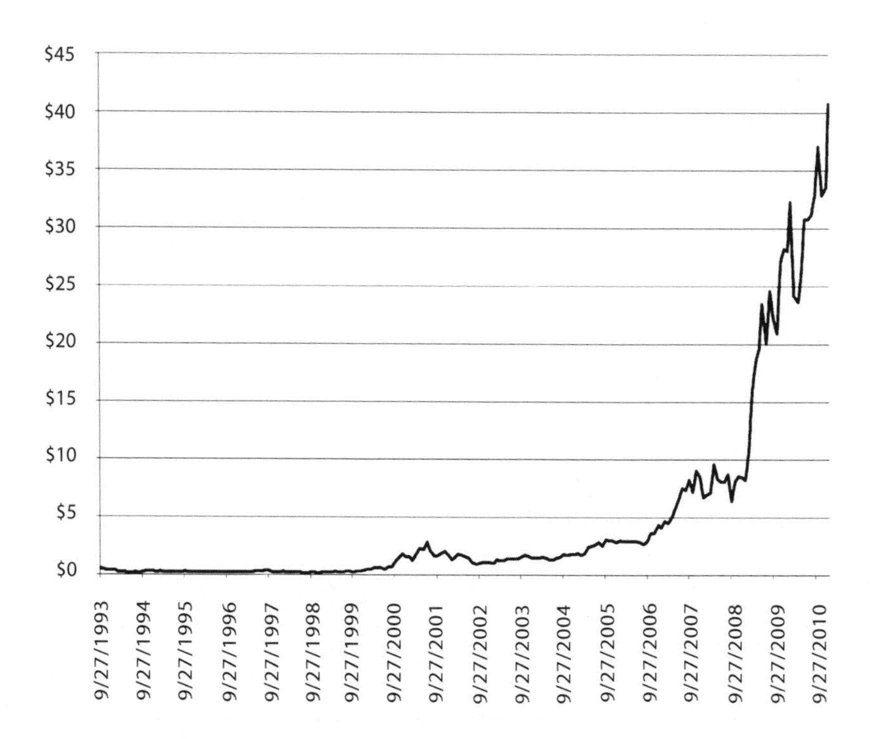

4.2 Green Mountain Coffee Roasters, Inc., Share Values (adjusted for splits)
Source: Graphic based on Yahoo! Finance (2011)

brands such as Newman's Own Organics. Third, by actively sourcing and promoting Fair Trade, Green Mountain would gain access to customers who look for the label. This turned out to be a very rewarding commitment. Figure 4.2 tracks the price of Green Mountain's stock through the beginning of 2011, showing steady and sometimes dramatic increases, although, of course, one cannot tell exactly how much of this is due to Fair Trade sales. In fiscal year 2009, Fair Trade Certified coffee represented more than 30 percent of the overall coffee Green Mountain shipped—a 36 percent increase in volume over the prior year and a record high for the company.[9]

Green Mountain's "integrated communication of its commitment to Fair Trade and organic certification, its philanthropic giving, its investment in coffee-growing communities, its Vermont roots, and its folksy, familial tone creates an unusually strong bond with customers. This relationship bundles together the company's concerns about living wage, environmental protection, and product quality and all these factors ultimately contribute to the Green Mountain brand" (Grodnik and Conroy 2007, 100). The extent to which others can follow Green Mountain's example and achieve similar success remains to be seen.

Strategy, Organization, and Engagement

Management scholar Valéry Bezençon and international business specialist Sam Blili (2009) sought to understand the strategies and managerial practices related to Fair Trade product distribution, with an eye toward analyzing whether or not these strategies and practices denote an engagement with Fair Trade tenets. What motivates retailers and distributors to offer Fair Trade products? How do they organize their businesses? Are they engaged with Fair Trade as a principle? Bezençon and Blili conducted their research in Switzerland, where Fair Trade products are now ubiquitous but certainly not the only comparable items on offer. The five companies they studied—Coop, Switcher, La Semeuse, McDonald's Switzerland, and Magasin du Monde—differ in size and in what they do, and therefore differ in terms of how they organize their Fair Trade product distribution.

Based on their study of corporate management reports, corporate social responsibility (CSR) reports, the companies' websites, and extensive interviews with employees responsible for managing Fair Trade or social and environmental products, Bezençon and Blili classified their cases along a scale of business engagement[10] with Fair Trade: from defensive CSR business to proactive CSR business to values-driven business to Fair Trade organization. For each case, the scholars also observed a primary meaning of Fair Trade within the company's organizational framework. For example, under pressure from Fair Trade activists, McDonald's Switzerland started offering a Fair Trade certified coffee. Bezençon and Blili call McDonald's a "defensive CSR business" and attach the meaning "contrast" because the company's Fair Trade purchases are in contrast to its usual sourcing practices.

The giant retailer Coop, a "proactive CSR business" already stocked environmentally and socially responsible products before adding Fair Trade to the mix. Coop's management saw Fair Trade as a coherent label that could help diversify the firm's offerings (meaning: "diversification"). Another "proactive CSR business," coffee roaster La Semeuse, adopted Fair Trade early on because of its owner-manager's ideology, not as a business strategy per se (meaning: "individual ideology"). The clothing manufacturer Switcher is deemed a "values-driven business" because the company paid attention to ethical trade long before it was able to offer items made of Fair Trade–certified cotton (meaning: "corporate culture"). Magasin du Monde, a "Fair Trade organization," takes the meaning "devotion" because the company exists to promote equitable commerce. At Magasin du Monde, selling Fair Trade products is a means toward the end of educating consumers about international trade issues.[11]

The point of this classification exercise is to show that very different motives and strategies can lead to progressive CSR policies and significant revenues for Fair Trade producers. Giants like McDonald's and Coop are not as dedicated to Fair Trade as are La Semuse, Switcher, or Magasin du Monde, but the magnitude of their Fair Trade purchases is far greater. However, the authors point out that large, mainstream distributors are not conveying a transformative message. They note that these firms' global business practices may be at odds with Fair Trade values and principles. What is clear is that mainstream companies are selling lots of Fair Trade products, while alternative businesses are selling the Fair Trade message.

Some scholars argue that mainstream demand for Fair Trade products is not a good thing, that large retailers are buying relatively small amounts of certified products but benefitting from "the image associated with the seal's values," and in turn weakening Fair Trade as a movement (Renard 20043, 93). This phenomenon has been labeled "McFair," (Jaffee 2007,

203), "clean-washing" (Low and Davenport 2005, 494), and "bluewashing" (the social and human-rights equivalent of "greenwashing").[12] Bezençon and Blili acknowledge the possibility that mainstream firms are sourcing Fair Trade products in order to enhance their images or (as in the McDonald's Switzerland case) to deflect negative publicity. But their work does not support the claim that mainstreaming is weakening the Fair Trade movement; rather, it is exposing a whole different set of customers to Fair Trade products while increasing revenues to farmers.[13]

Fair Trade and Private Labels

Development scholars Stephanie Barrientos and Sally Smith studied Fair Trade house brands in major UK supermarkets. Their angle is especially interesting because, since they outsource the products' packaging and labeling, supermarkets do not need to be Fairtrade International licensees to sell house-brand Fair Trade lines. "Since supermarkets are therefore not necessarily bound by Fair Trade rules and regulations, their suppliers are potentially being exposed to the types of practices and pressures that exist in conventional production networks" (2007, 103).

Barrientos and Smith provide data about eight major UK supermarkets, the top four of which account for almost three-fourths of the nation's grocery sales. Fair Trade products are nothing new to shoppers in these supermarkets, but in 2000 the Co-op chain did something different: it introduced the first ever house-brand Fair Trade Certified product, a milk chocolate bar. Since then the company has converted *all* of its house-brand coffee and chocolate products to Fair Trade. Currently, Co-op vies with Tesco, the UK's leading supermarket, for the top spot in terms of Fair Trade sales volume. Co-op offers by far the largest selection of Fair Trade products and has the most Fair Trade house brands (54 percent of the 123 Fair Trade products stocked in 2005), but Tesco is a much larger company. All of the top eight supermarkets offer multiple Fair Trade selections and house-brand Fair Trade products.

Why have these mainstream companies aggressively sourced and marketed Fair Trade products to which they can attach a private label? Citing an interview with Tesco's Fair Trade product manager, Barrientos and Smith identify three reasons: "customer demand, commercial opportunity, and brand value." Statements about supporting marginalized producers or addressing inequities in global trade are conspicuously absent in this rationale. Even at Co-op, a company with very progressive CSR policies, a CSR manager told the researchers that "Fair Trade is part of the responsible retailing brand that we are trying to develop as a business strategy" (2007, 106–7). On one hand, this statement shows that British shoppers are demanding Fair Trade products. On the other hand, it means that supermarkets are driven to search for Fair Trade sources instead of waiting for Fair Trade products to be offered to them. Indeed, it appears that in the United Kingdom there is no need for Fair Traders to "target" businesses.

Because of the power dynamics involved, Barrientos and Smith approach this development with skepticism. For example, in order to meet fluctuating demand, supermarkets might switch between Fair Trade suppliers as they do with other suppliers, or pressure producers to enter into contracts whereby they supply on a "just-in-time" basis, forcing them to ship their produce in response to volatile consumer demand. Both of these possibilities violate the Fair Trade standard of long-term, contractual arrangements between producers and buyers—the first in fact, and the second in principle. Still, the authors acknowledge another possibility: supermarkets' incorporating Fair Trade into their house-brand retailing could challenge their conventional buying practices. After all, there would be a serious credibility price to pay if a grocery chain were accused of exploiting Fair Trade producers.

Barrientos and Smith proffer two case studies of producer-buyer relationships in supermarket house brand–driven value chains: cocoa from Kuapa Kokoo in Ghana, and fresh fruit from Thandi in South Africa. In the cocoa case, the Co-op chain worked with Day Chocolate to create their private-label brand, and Kuapa Kokoo worked with a German processor who committed to handling the large quantities of cocoa required to meet Co-op's needs—key because "running out of all Co-op chocolate on store shelves would be a commercial disaster" (2007, 112). The case of Thandi fruit is remarkably different because there was no intermediary as there was with Day Chocolate, fresh fruit is perishable, and fruit exported to Europe must meet exacting food safety and cosmetic standards. Here, Fair Trade and conventional value chains look extremely similar. For example, in a preseason marketing trip, a Fair Trade producer established contracts with five UK supermarkets. But when the fruit arrived on time in the United Kingdom, only two supermarkets honored their contracts, and one of these waited four weeks and then accepted a delivery smaller than what was formerly agreed upon. The fruit's quality diminished in the meantime, and the supermarket imposed a "poor quality fine" on the producer group (cf. Aparicio, Ortiz, and Tadeo 2009; Freidberg 2004).

Such problems are neither ubiquitous nor rare. In revealing them, Barrientos and Smith do not mean to imply that South African workers have nothing to gain from Fair Trade. Rather, they emphasize the imperative to adhere to Fair Trade standards even if supermarkets that develop Fair Trade house brands are not strictly required to do so. They also note that there is almost no direct contact between supermarket buyers and Fair Trade producers, and that there are no incentives for grocery chains to work with smaller producer groups that may require support (e.g., prefinancing or technical assistance). This raises a further question, addressed in the following section: Will competition drive out businesses that most closely adhere to Fair Trade principles?

Alliances and Networks

Given the enormous reach of UK supermarkets that stock Fair Trade products, how do small, values-driven Fair Trade companies compete? Marketing scholar Iain Davies's study draws attention to the value of interorganizational networks. He posits that networks are important because they "form the basis of competitive positioning of many modern Fair Trade companies." They facilitate competitive, intellectual, and ideological development (2009, 2).

Davies conducted participant-observation and interview-based research in three UK companies with explicit Fair Trade agendas: Equal Exchange, Cafédirect, and Day Chocolate (table 4.1). Because at times conflict exists between companies' sales/profit objectives and their ideological purposes, the inquiry focused on how networks help resolve such conflict. Networks include other businesses, NGOs, journalists (for press coverage), and multinational corporations (to gain grocery store shelf space).

Davies identified key *networks of ownership*, *networked supply chains*, and *distribution networks*. In networks of ownership, the benefit is shared competencies. For example, four organizations have contributed to Cafédirect's success: Equal Exchange (marketing), Twin Trading (supply chain management), Oxfam (campaigners), and Traidcraft (representation within

Table 4.1
Davies's Cases: Fair Trade Companies in the U.K.

Company	Ideology
Equal Exchange	Demonstrate alternative trading through relationships
Cafédirect	Pioneer Fair Trade into mainstream to maximize sales and grow income
Day Chocolate	Demonstrate successful alternative trading to drive change in the cocoa industry and pass profits back to growers

Source: Davies 2008, 5

alternative trade networks). Networked supply chains have helped these companies focus on marketing and logistics. By and large, they work with importers, processors, and packagers because it would be too costly to take on these activities themselves. Networks of distribution and retailing link the companies to wholesalers and specialist distributors. In the case of Equal Exchange, even the products they sell through mail order are delivered to the company headquarters by a wholesaler. Cafédirect partnered with a specialist distributor to get vending-machine and coffee-shop distribution.

These networks allow Fair Trade companies to be *virtually* integrated and to appear much larger than they actually are. Networks also make it possible for companies to improve their competitiveness by pooling information (instead of each one employing specialists). Some network members are nontrading partners. For instance, Equal Exchange UK keeps up a relationship with Equal Exchange America, even though their only historical link is that the former borrowed the latter's name in the 1980s. Employees of both companies visit each other, share ideas, and have organized joint visits to coffee cooperatives in Mexico. Day Chocolate partners with the NGO Christian Aid, whose members organize chocolate sales events. The NGO has also helped Day with publicity and marketing. All three case companies nurture further nontrading relationships through their membership and participation in groups such as the European Fair Trade Association.

What makes for successful networks? For the businesses Davies studied, shared core values are especially important, because organizations founded on similar principles will be more cooperative, trusting, and trustworthy in their dealings with each other. These companies' networks form the basis for both ideological perpetuation and business growth. This study suggests that strategic networking helps smaller, value-driven firms retain and increase their market share even as more mainstream businesses are sourcing Fair Trade Certified products.

Summary: Businesses and Fair Trade

Companies operate in markets, and if there is market demand for social and environmental sustainability, they will adopt progressive CSR policies. They may also adopt CSR to facilitate price discrimination, since some consumers are willing to pay more if they believe a company is socially responsible (Mohan 2010). Of the many companies the "business" studies scrutinize, only McDonald's appears to be using Fair Trade as a defensive strategy in response to negative publicity about the firm's sourcing practices. La Semuse and Switcher were early adopters of Fair Trade because it clearly fits their CSR policies. The rest are integrating Fair Trade into some form of cause-related marketing. Green Mountain, Coop, and all eight UK supermarkets are using Fair Trade as a brand extension. They are responding to consumers' demands while attempting to gain or keep their customized strategies of differentiation (e.g., Green Mountain's "good business" image, and supermarkets' Fair Trade house brands). For Magasin du Monde, Day Chocolate, Equal Exchange, and Cafédirect, Fair Trade is a core business strategy. This latter group of businesses, all of which stake their reputations on ethical trade, creatively uses networks to help them compete with larger companies, get a foothold in new niches, and protect their market share.

Large mainstream businesses are now buying and selling vastly more Fair Trade products than are small, 100 percent Fair Trade companies. Fair Traders are understandably concerned that corporate appropriation of Fair Trade will result in a watering down of Fair Trade's transformative message. Some of the studies examined in this chapter (notably Bezençon and Blili [2009] and Barrientos and Smith [2007]) could be read to suggest a review and revision of the Fair Trade standards in light of rising demand from large corporations. For example, such companies could be required to promote Fair Trade via consumer education rather than simply source Fair Trade products.

We have also seen several examples of successful companies that, despite their modest size, are consistently selling a message of ethical trade and social justice alongside Fair Trade products. It does not appear that the "mainstreaming" of Fair Trade products is hurting these firms or forcing them to dilute their message in order to be competitive. In fact, it is possible that networks of such businesses and their allies could in time create what has been called an "Alternative High Street"—an ethical space in which social justice, human and animal rights, and environmental welfare come before profits, and products sourced according to these values "do not simply constitute one choice amongst many, but instead they supplement the individualized approach of shopping for a better world being pursued elsewhere" (Low and Davenport 2009, 9).

TARGETING CONSUMERS

Consumer campaigns aim to stir up demand, and education is the first step, since consumers cannot ask for things they are unaware of. Fair Traders' twofold goal has been to teach consumers that they possess the power to make a positive difference in the developing world through their purchasing behavior and to encourage them to *act* on this knowledge. This educational outreach began through the isolation of target consumers: those who are most likely to be socially conscious purchasers. For example, TransFair USA's 2002 annual report identified the potential market for Fair Trade coffee as individuals who are between twenty-five and forty-five years old, have an annual income of $35,000 or more, are college educated, live in "liberal" towns (particularly in the Northeast, Midwest, and Northwest) and are not put off by high prices. Similarly, European Fair Traders first tried to target the likeliest customers. Now, Fair Traders on both sides of the Atlantic have extended this focus beyond a socially aware consumer base, seeking to attract and respond to higher-volume mainstream demand.

European labelers were the first to try to graft Fair Trade onto the everyday activities and interests of more mainstream consumers. For example, the UK-based Fairtrade Foundation produced educational packs suitable for children of various ages and adaptable to a range of curriculum areas. Max Havelaar Netherlands organized a writing contest for Dutch youth, who wrote essays discussing the international development of Fair Trade. Max Havelaar Belgium published a comic book for classroom use,

The Struggle for the Black Bean, and a board game aimed at demystifying the global economy, the Max Trade Game. Max Havelaar France developed a play entitled "Aye Aye Aye Café," an adaptation of the (fictional) history of the late-nineteenth-century hero Max Havelaar, who tried to battle layers of Dutch-colonial corruption in Indonesia. Several of the Max Havelaar labelers reached out to their Fair Trade converts, asking them to become "ambassadors of the elephant" (Max Havelaar's original trademark; Linton, Liou, and Shaw 2004).

Today, arguably the most well-known happening around Fair Trade is the yearly UK "Fairtrade Fortnight"—a family-centered series of events that in 2009 included a sports day, a chance to "go bananas" by eating a Fair Trade banana within a specified twenty-four-hour period, and awards for the best locally organized activities to increase awareness of Fair Trade and consumption of Fair Trade products (Fairtrade Foundation 2009b). Max Havelaar France coordinates a like-minded observance, *la quinzaine de commerce équitable,* marked by Fair Trade brunch parties and events featuring celebrities and government officials (Max Havelaar France 2009). In Germany, "World Shops," supermarkets, chefs and restaurants, and a diverse community of activist organizations sponsor Faire Woche (Fair Trade Week), an event aimed at raising public awareness of global trade issues (Fair Trade Deutschland 2011).

When does awareness translate into action? In their study of Fair Trade consumerism, sociologists Matthias Varul and Dana Wilson-Kovacs (2008) identify two major forms of moral self-identity: *taste for ethics* and *ethical taste.* "Taste for ethics" describes truly committed individuals who actively and consistently search for the ethically best buy, whereas "ethical taste" melds the moral and aesthetic judgments of more passive shoppers who are pleased when a product tastes good and *is* good is some ethical sense.[14] What of the majority of current or potential Fair Trade consumers, people who—perhaps due to Fair Traders' educational efforts—have developed ethical taste? Geographers Raymond Bryant and Michael Goodman's study of Fair Trade consumers in the United Kingdom offers this example from the Fairtrade Foundation's 2003 Internet homepage: "My name is Anna. I love great coffee and I *trust* the FAIRTRADE Mark. When I see the FAIRTRADE Mark I know my coffee is made with *quality* beans brought directly from farmers around the world at a fair price. It's a partnership for a better future" (2004, 357, italics in original).

Bryant and Goodman identify key elements of a "solidarity-seeking commodity culture" in which consumers purchase "ethical commodities" and *the act of consumption is politicized* through "materially and socially embedded ethical relationships," that is, ways that people relate to Fair Trade brands and those who promote them. Creating such a culture requires "detailed dissemination of information to Northern consumers saying what Fair Trade is and why it is needed" (2004, 358, italics added). Such knowledge can be quite place-specific—for example, giving potential consumers information about the region, community, and producer group that a product comes from. In the example of Anna above, successful marketing has brought her to connect (in her mind and at the store) the coffee she enjoys, the FAIRTRADE Mark, and a fair price for farmers.

Consumer-Based Research

Marketing scholars and social scientists have quite a bit to say about why people might want to consider the livelihoods of faraway farmers as they buy coffee, tea, and bananas and as they make their grocery lists. Researchers typically begin with the logic that "knowledge or beliefs lead to general attitudes that in turn lead to intentions and behavior" (de Pelsmacker and Janssens 2007, 363). Once a consumer perceives a problem (e.g., clothes made in sweatshops or coffee farmers not earning enough to survive), she or he thinks of possible ethical alternatives. To evaluate the alternatives, the shopper may consider their strengths and weaknesses in order to choose a "best" alternative, or generally assess the effectiveness of market-based fixes. The first construct focuses on solving a problem via ethical consumption, the second is more skeptical about ethical shopping as a way to correct abusive practices or global market inequalities (Hunt and Vitell 1993). "Control beliefs"—how much a person thinks that his or her actions make a difference—also influence what does or does not go into the grocery cart (Shaw and Shiu 2003).

While it is clear that consumers' beliefs, perceptions, and attitudes influence their behavior, it is also well documented that these elements *alone* are poor predictors of behavior.[15] Because of this, the studies reviewed here seek to reveal factors that effectively link values and perceptions with actual or hypothetical behaviors. Some of them also address the fact that often, although by no means always, the Fair Trade label comes with a price premium that, in effect, asks shoppers to pay more for a guarantee that the producer received a fair price and is farming in an environmentally sustainable way. There is a large body of evidence to show that people will pay extra for socially labeled goods,[16] but it is unclear how much more various groups of consumers will pay, and for what sort of label.

When Do Shoppers Link Their Consciences with Their Wallets?

At the core of many social scientists' research on consumers and Fair Trade is an effort to better understand how knowing about Fair Trade influences (or fails to influence) peoples' actual shopping habits. Interviewing customers in World Shops (businesses that specialize in ethically traded goods) throughout Italy, economists Leonardo Becchetti and Furio Rosati (2007) found that the amount shoppers spent on Fair Trade products was closely linked to their *knowledge* about Fair Trade and other aspects of socially responsible consumption. Interviewees expressed awareness of and concern about the following (in order of importance): fair prices for farmers, working and living conditions, child labor, environmental conservation, transparency, local public goods, prefinancing, price stabilization, and long-term relationships between producers and buyers.[17] The length of time a person had been a World Shop customer also affected Fair Trade expenditures through awareness, confirming these businesses' success as promoters of socially responsible consumption.[18]

Economists Patrick De Pelsmacker and Wim Janssens (2007) developed a model of Fair Trade buying behavior that incorporates perception of the quantity and quality of information about Fair Trade and attitudes toward Fair Trade products. They were especially interested in knowing how much the *information people receive* about Fair Trade and their *perceptions of the products* influence whether or not they buy.

De Pelsmacker and Janssens's focus-group and survey research in Belgium indicates that, besides concern about North-South trade issues and knowledge about Fair Trade, specific interest in Fair Trade products, perceived quality and quantity of Fair Trade information, price acceptability, and product likeability are positively related to buying Fair Trade products. Interestingly, with these factors taken into account, shopping convenience does not significantly affect peoples' buying behavior. It appears that becoming interested *in the actual products* transforms an abstract interest in the issues that Fair Trade seeks to address into actual Fair Trade buying behavior.[19] This suggests that the most important obstacle for Fair Traders to overcome is people's indifference toward Fair Trade coffee, tea, and so on.

How important is price? Economist Chris Arnot and his colleagues (Arnot, Boxall, and Cash 2006) conducted their research in a coffee shop on the campus of a Canadian university. The shop offered two varieties of medium-roast coffee: a Columbian (the best seller) and a Fair Trade coffee from Nicaragua that tastes very similar to the Columbian. To facilitate the study, the shop's manager discounted each coffee from ten cents to fifty cents per cup on a rotating basis and allowed researchers to conduct short surveys of coffee customers. The resulting analysis of nearly 500 surveys suggests that Fair Trade consumers are not particularly price-sensitive because their choice of Fair Trade coffee remained consistent even when the (non–Fair Trade) Columbian coffee was discounted. On the other hand, consumers who usually purchased the Columbian coffee were likely to switch to the Fair Trade option when it was cheaper.

Economists Arnab Basu and Robert Hicks were also interested in how the attributes of Fair Trade products influence whether or not people buy them and how much they will spend. In addition, they explored the role of cultural context in the attributes-purchasing relationship. College-student research subjects first read the following:

> WHAT IS FAIR TRADE?
> Advocates argue that Fair Trade certified products ensure that farmers, workers, and artisans are paid a fair price for their products or labor, do not use child labor or forced labor, have healthy and safe working conditions, use sustainable and environmentally friendly production methods, and have long-term and direct relationships with producers and buyers. Others feel that fair trade is discriminatory against producers who are not eligible for the program and countries that do not have the resources to institute a Fair Trade program. (2008, 9)

The researchers then offered the subjects an opportunity to choose among coffees with various (hypothetical) "performance labels" and nonlabeled coffees, at different price points (figure 4.3 shows some of these labels). The performance labels include statements about how much the coffee's Fair Trade sales have increased farmers' incomes and about the growth rate of

Assume that you are going to buy a cup of coffee. If you could only choose from the following three choices, which one would you choose?
(Please check only one of the boxes at the bottom of this page).

Coffee A

$2.50

FAIR TRADE CERTIFIED®

This Brand's Fair Trade Performance:
- Increased Grower Revenue: 50%
- Increased Grower Participation: 75%

Certified Fair Trade by the Costa Rican Coffee Growers' Association.

Grown in Costa Rica

☐ **Choose Coffee A**

Coffee B

$2.50

FAIR TRADE CERTIFIED®

This Brand's Fair Trade Performance:
- Increased Grower Revenue: 75%
- Increased Grower Participation: 25%

Certified Fair Trade by the United States Department of Agriculture.

Grown in Colombia

☐ **Choose Coffee B**

Coffee C

$1.00

No Information Available

Grown in Brazil

☐ **Choose Coffee C**

☐ **I wouldn't buy any of these**

4.3 Basu and Hicks's Questionnaire

"participation," which could be read as new members joining the producer group the coffee came from or new Fair Trade coffee cooperatives being established in the same producing country. Such information could help coffee drinkers evaluate the degree to which buying the labeled product is actually lifting farmers out of poverty. There were multiple combinations of Fair Trade, organic, and "certified by [the country of origin's] coffee association" labels, with varied performance attributes.

In both the United States and Germany, the Fair Trade label had the strongest effect on willingness to pay a higher price as long as farmers' hypothetical income did not increase "too much." The study design led participants to consider "the interplay between poverty and inequality aversion as participating growers in a Fair Trade program becomes better-off" (Basu and Hicks 2008, 15). Respondents in the United States accepted up to 75 percent increases in farmer revenues, whereas for German respondents the limit was 55 percent. In other words, Americans were willing to tolerate more (hypothetical) inequality between Fair Trade farmers and their nonorganized counterparts. The results of this experiment indicate that information-rich performance labels could reward producers by increasing consumers' willingness to buy Fair Trade products and to pay higher prices, and that in order to induce consumers to buy Fair Trade products and help change norms about what we buy and why we buy it, packaging should communicate more than a Fair Trade logo. However, performance labels will work differently in different markets. Other researchers have also demonstrated that Fair Trade products' packages and promotional materials can further a sense of connection between consumers and producers. Besides increasing Fair Trade sales, packaging that offers specific information about the people who grew the product encourages consumers' engagement with the underlying issues (Goodman 2004; Ozcaglar-Toulouse, Shiu, and Shaw 2006).

A Deeper Look at Values

Management scholar Carolyn Doran sought to discover the values most salient to Americans who are already Fair Trade consumers. She surveyed Internet shoppers on the websites of Fair Trade retailers Guayaki, Two Hands World Shop, Peace Coffee, and Just Coffee. Doran employed the Schwartz Value Survey comprised of ten "value types": universalism, benevolence, conformity, tradition, security, power, achievement, hedonism, stimulation, and self-direction (2009, 2). Loyal (as opposed to intermittent) Fair Trade shoppers scored high on self-direction, a value set that could promote breaking from convention and paying higher prices for products that are often harder to find. But, above all, loyal Fair Trade customers esteemed universalism. This value set includes unity with nature, a world of beauty, and protecting the environment. In addition, loyal Fair Trade consumers ranked benevolence values lower than intermittent consumers did. Both universalism and benevolence values have to do with supporting others, but universalism values encompass *all people and nature*, whereas benevolence values focus on *the in-group*. It appears that, as consumers become more involved with Fair Trade as a means of consumption, they become less involved with supporting the in-group. Recall that the same, with nuances, is true of Fair Trade producers (see chapters 2 and 3). Groups of producers who are highly involved in Fair Trade as a means of production usually use the Fair Trade premium in ways that benefit entire communities. This suggests that commitment to Fair Trade values (as well as products or markets) is related to a more expansive sense of membership—seeing one's connection to people everywhere, and to the environment.

Marketing scholar Nil Ozcaglar-Toulouse and his colleagues Edward Shiu and Dierdre Shaw (2006) also studied "concerned consumers": French people who visited Fair Trade websites and readers of *Nouveau Consommateur* magazine (a publication oriented to the Slow Food movement). Among them, the regular Fair Trade purchasers viewed their Fair Trade consumption as a personal norm because they felt an ethical obligation toward the people who produce their food. Since concerned consumers who currently do not buy Fair Trade products are potential new buyers, the researchers focus on ways to influence their sense of ethical obligation.[20] Citing a study conducted in the United Kingdom, Ozcaglar-Toulouse, Shiu, and Shaw note changes in consumer behavior *to the extent that ethical concerns become part of everyday conversation* (2006). Becchetti and Rosati's (2007) interviews with World Shop customers also indicated that membership in a secular or religious volunteer organization is positively related to knowledge about Fair Trade and awareness of the issues it seeks to address, and that people who know more about Fair Trade spend more money on Fair Trade products. It follows that Fair Traders should target peer and social groups with the goal of creating new communities of concerned consumers who display their membership by purchasing Fair Trade products.

Ozcaglar-Toulouse, Shiu, and Shaw also encourage marketing Fair Trade as part of a bigger picture of social justice and environmental sustainability by connecting it with other issues, such as organic farming, supporting local farmers, food safety, and animal welfare. Acknowledging Fair Trade consumption as an interconnected activity allows for a brand image that can reach consumers with many different—and not necessarily related—concerns.

Fair Trade as a Human Right

To date only one study considers both attitudes about human rights and willingness to pay more for fairly traded goods. Political scientist Shareen Hertel and her colleagues Lyle Scruggs and C. Patrick Heidkamp conducted a nationally representative phone survey in the United States in which they asked questions about freedom from torture, freedom of thought and expression, and *the right to a minimum standard of living*. Over 60 percent said that the right to a minimum standard of living was inviolable. Then the researchers asked several questions about willingness to pay more—and how much (if any) more—for a sweatshop-free sweater or a pound of Fair Trade coffee.[21] The study's results show a clear association between believing in a right to a minimum standard of living and willingness to pay at least a dollar per pound more for Fair Trade coffee.[22] Even among the 381 respondents who said they had never seen a Fair Trade label, 26 percent expressed willingness to pay at least a dollar more for it.[23] Hertel, Scruggs, and Heidkamp conclude that "American consumers are more likely to 'put their money where their values are' in progressive defense of economic rights than has heretofore been suggested" (2009, 457). This implies that there is untapped market demand for ethically produced goods.

Summary: Consumers and Fair Trade

The studies summarized in this section do not all start with the same questions, but there is remarkable consensus in their findings. The research about knowledge, attitudes, and beliefs focuses on how these are related to consumption of Fair Trade products and willingness to pay more for them. Basu and Hicks's work indicates that—compared to organic- and shade-labeled coffee—consumers are willing to pay more for Fair Trade.

De Pelsmacker and Janssens show the importance of generating interest in actual Fair Trade products, not simply in Fair Trade as a movement or fair trade as a concept. Both Becchetti and Rosati's study of World Shop customers and Arnot, Boxall, and Cash's coffee shop experiment indicate that income is not a determinant of Fair Trade consumption. Rather, concern about global inequalities and specific knowledge about socially responsible purchasing options are what motivated World Shop customers and college students to choose Fair Trade. Doran's "values" survey also evidences a link between people's involvement in "shopping for change" (e.g., by buying Fair Trade products) and their sense of membership in a global community.

Most of these studies intentionally focus on consumers who exhibit some interest in ethical consumption. They highlight the gap between interest and behavior and suggest ways to reduce it by (1) targeting peer and social groups so that ethical concerns play a role in people's daily interactions, (2) designing packaging that, besides telling about the product, offers information that promotes a sense of connection with producers, and (3) connecting Fair Trade to a bigger picture of ethical trade, environmental sustainability, and human rights.

CONCLUSION: FAIR TRADE AS POLITICAL CONSUMPTION

Collectively, this research suggests that people who regularly buy Fair Trade products are linking their global concerns with their shopping habits; they view their purchasing as a form of political consumerism that is of course facilitated by businesses that promote Fair Trade. Political scientist Dietlind Stolle and his colleagues Marc Hooghe and Michele Micheletti found that, rather than crowding out other forms of political participation, political consumerism is "part of an array of activist performances that serve to broaden the spectrum of politics" (2005, 260). Their study of Canadian, Belgian, and Swedish students shows that, while political consumers vote and contact their elected representatives at about the same rate as nonpolitical consumers, the former have a higher sense of their own political efficacy and a greater belief in the effectiveness of unconventional political participation. Studying Danish consumers, sociologists Jørgen Goul Andersen and Mette Tobiasen (2006) found exactly the same thing. For political consumers, voting with one's wallet *augments* other forms of political involvement. This is not to say that everyone who buys a Fair Trade product now and then is a political consumer, but these studies indicate that by connecting people's concerns to ways we can vote with our money, successful Fair Trade marketing does in effect politicize consumption (see also Goodman and DuPuis 2002).

Along these lines, geographer Nick Clarke and his colleagues argue that Fair Trade should be "understood as a political phenomenon, which through the mediating action of organizations, coalitions, and campaigns, make claims on states, corporations, and international institutions" (2007, 585–86). Their point is that ethical consumption campaigns do not really seek to mobilize individuals as *individualistic* consumers (which is not surprising, given that consumer research never finds links between individualism and ethical consumption). Rather, Fair Traders employ a whole

> spectrum of actions through which consumption is problematized as both an object and medium of ethical commitment and political participation. This spectrum ranges from individualized, discrete activities such as purchasing fairly traded products in the supermarket to more sociable practices such as involvement in local campaigns to have schools, universities, or towns certified as "fair trade" through to explicitly political engagement, whether through individualized petition signing or collective involvement in mass demonstrations. Even the most individualized and consumerist of these activities is, then, *connected to a broader range of actions*. (2007, 587, italics added)

Clarke and his colleagues point out that it is only possible to act as a Fair Trade consumer because Fair Traders campaign, educate, trade, and market. To illustrate, they focus on the UK-based Christian organization Traidcraft, "an intermediary actor in networks of civic activism and political participation" (591). Traidcraft's activities fall into three categories: "individualistic" activism (e.g., wearing a campaign button or buying Fair Trade coffee), "contact" activism (e.g., writing to a lawmaker or signing a petition), and "collective activism" (e.g., participating in a demonstration or joining a union). Each of these organizational dimensions translates into a simple mode of personal action: "you buy—and we can trade; you donate—and we can support; you campaign—and we can influence" (591). Instead of trying to change individuals' opinions and preferences, Traidcraft seeks to extend peoples' existing dispositions into new areas. "We used to try and lead people to think and act in particular ways. Now we try and respond to people and provide outlets for their energy and commitments" (592).

There is no single US organization that mirrors Traidcraft but rather a network of Fair Trade businesses and activist groups—some faith-based—that as a whole functions in much the same way (though at some nodes in the network the holistic picture may be obscure). In the next chapter we will meet some of the individuals who comprise this network and learn how they are expressing their "ethical taste."

Chapter 5

Fair Trade Activists in the United States

Rebecca Kahn and April Linton

Some people's commitment to and involvement with Fair Trade extends well beyond their own shopping habits. What motivates them to become activists? How do they understand the producer-consumer relationship? The individuals profiled here have a vested interest in ethical trade, although not all of their activism focuses on promoting Fair Trade Certified products. Many are affiliated with Fair Trade Towns, a networked initiative that unites community activists, businesses, faith-based organizations, and educational institutions with the goals of growing Fair Trade and supporting local businesses and farmers. They represent varied organizations, with diverse connections to the Fair Trade movement.

This chapter highlights eight in-depth interviews conducted in October 2008 in New York, Massachusetts, Washington, and Vermont.[1] The initial interviewees were Fair Trade Towns committee chairs; these in turn provided other contacts. This allowed us to interview people from varied organizations with diverse connections to the Fair Trade movement. We employed a schedule of questions to guide the semistructured interviews but allowed the conversations to follow a natural direction, partially led by the interviewee. The general themes of each interview were defining Fair Trade and its place in the larger ethical trade and conscious-consumption

Table 5.1
Interview Subjects

Name	Connection to Fair Trade
Taylor Mork	Farmer Representative, Crop to Cup Coffee. Imports coffee from small holder farms with high ethical standards, but does not have certification.
Ron Zisa	Buys ethically traded (certified and noncertified) coffee for 12,000+ member urban food co-operative market.
Yuri Friman	Fair Trade activist, member of Fair Trade Towns USA.
Barth Anderson	Specialty coffee roaster (Barrington Roasters), has both certified and noncertified fair trade coffee.
Alexandra Mello	Fair Trade activist, member of Fair Trade Towns USA.
Judith Belasco	Works in the area of ethical food outreach and education, incorporates fair trade products into her work and home environments.
Pattie Cippi Harte	Brings Fair Trade products to her office, Jewish Community Centers of America.
Stephanie Celt	Director of a Washington State Fair Trade coalition, primarily works toward changing governmental trade policies.

philosophies and the interviewee's relationship with these broad topics. To add a visual dimension to the research, we asked participants to make a drawing of how they envision producer-consumer relationships within Fair Trade, and then to describe their drawing. The drawings offered a useful avenue for interviewees to express their identities and describe their relationships with others, including farmers, within the Fair Trade movement.

LOCAL AND GLOBAL: RESTRUCTURING INTERPERSONAL RELATIONSHIPS

> More than abstinence from harming others . . . a society is viewed as just insofar as it provides a structure of interpersonal relationships, incentives, and reinforcements to virtue. (Lacy 2000, 9)

Our interviewees' different points of entry into the world of ethical trade illustrate the reasons for various understandings of and current relationships with the movement. One person attributed his connection to his activist days in the 1960s; another told a story of labor atrocities she witnessed on US banana plantations; yet another gave examples from his background in agronomy to justify his personal beliefs and connections surrounding ethical trade in the specialty coffee industry. One woman who had spent over a year living in a community that grows some Fair Trade Certified coffee and tea cited her particular connection with people in India: "I feel I know them, I know [their] culture, because I lived there."[2]

How and why are interpersonal relationships structured on the basis of economic processes and consumption practices? Fair Traders are globalizing their interpersonal relationships—actually and ideologically. In the process of trying to change economic and consumer relationships, they mark their place in their own communities, both local and global, by "calling for a realignment of human social interaction in the context of place and food" (Feagan 2007, 33). They are doing so via communities and interpersonal relationships structured around changing the way trade is carried out.

Communities

Communities do not have to be geographically bounded; they can represent ideologies that connect geographically and socially distinct individuals or groups (DeChaine 2005, Feagan 2007; see Lacy 2000 for comparison). The communication of people who have mutual concerns about ethical trade creates ideological communities. Through "buy local" campaigns, community supported agriculture programs, and Fair Trade, people are attempting to deepen bonds with the farmers who produce their food. Members of the aforementioned communities are helping others in the global North become aware of and connected to the people along the value chains that their coffee, tea, and chocolate follow—the people who add value to a product by harvesting, cleaning, drying, sorting, packing, and roasting it. This is an effort to promote *global citizenship*: seeing oneself as part of a global community and understanding the reciprocal relationships between geographically and culturally separated groups of people. At the same time, local relationships are strengthened. Fair Trade communities develop and grow through networks such as those facilitated by Fair Trade Towns USA, state-level coalitions, faith-based organizations, and consumer-awareness groups.

Community Economies

In an ideal Fair Trade value chain, the grower, exporter, importer, distributor, retailer, and end consumer all work together in a transparent and respectful relationship; they create a community economy. The goal of a community economy is to produce a knowledge that strengthens the social economy and helps to build it over time, enlarging its creativity, capacity, and credibility.

The eight interviewees that primarily inform this chapter are spokespeople for distinct microcommunity economies and are aware of their connection to the other actors in their particular value chain. They promote the community economy over individuals' interests. For example, specialty coffee roaster Barth Anderson talked about "exploring different ways to do good sustainable business, sourcing excellent coffee."[3] He illustrated three farm-to-cup pathways (including Fair Trade) that he finds acceptable because there are relatively few steps between the farm and the end consumer, and one that he does not support, the "traditional commercial" value chain.

Recognizing Interdependence within Communities

Fair Trade represents a complex network of relationships. The roaster quoted above alludes to his customers' economic and social relationships with him and his colleagues, and his own economic and social relationships with coffee farmers. The farmers and coffee pickers must understand the land and terrain around them in order to produce a quality coffee that will be enjoyed in Anderson's local community. When people recognize this web of relationships in their consumption choices, they are "encouraged to reevaluate their community, their food system and their role in it" (Cooley and Lass 1998, 228).

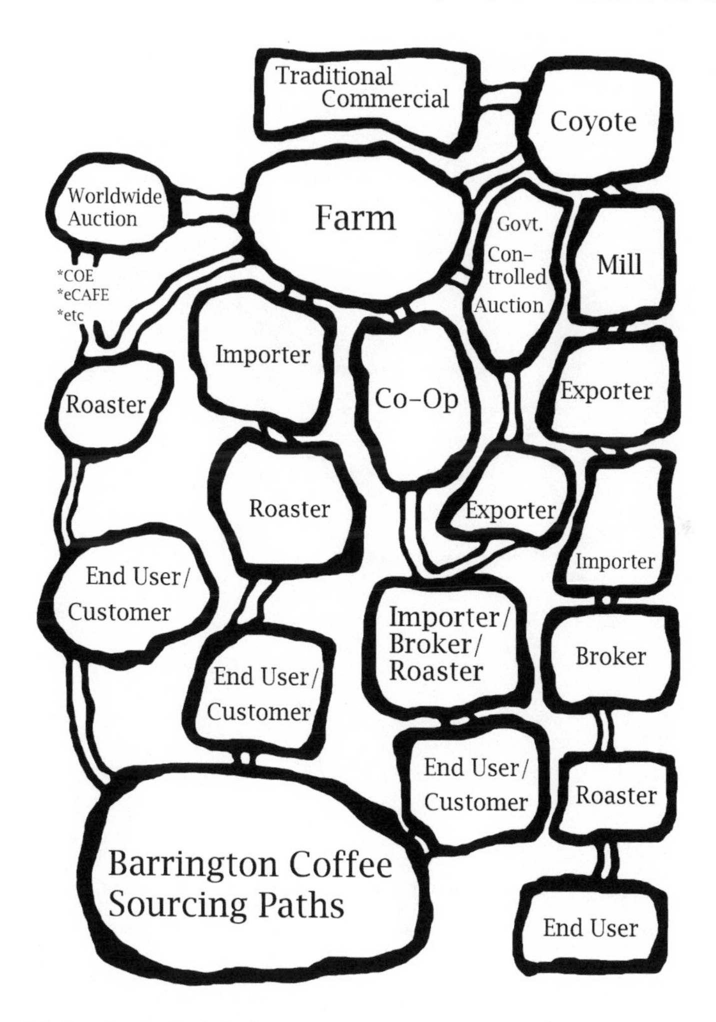

5.1 Drawing by Barth Anderson

Recognizing the relationships that exist within community economies is merely the first step in this process of reevaluation. Taylor Mork, who refers to himself as a farmer representative in his business of importing, roasting, and selling coffee, believes that the power of these relationships lies in the level of transparency that is present and acknowledged. He posits that it is possible and beneficial to have open, connected coffee-trading relationships, although historically the global coffee trade has been dominated by the consuming parties.[4]

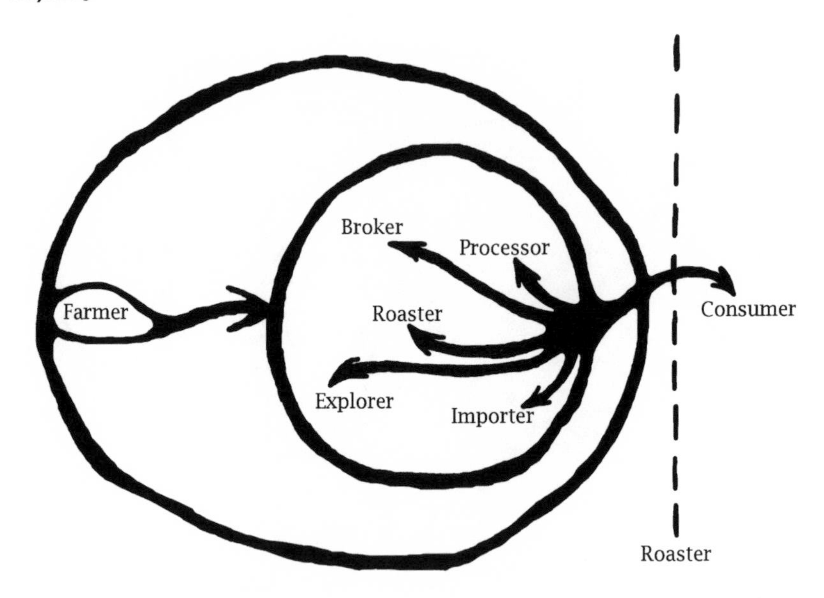

5.2 Drawing by Taylor Mork

Interpersonal Relationships

For over fifty years, most Americans have lived in cities or suburbs. Today only about 20 percent of the population inhabits areas the US Census designates as "rural"; many consumers have become ignorant of the places where food comes from and the people who grow it. Recent trends such as Community Supported Agriculture (CSA) programs, "buy local" campaigns, and Fair Traders' efforts to link farmers and consumers all speak to a renaissance of human interaction in consumption.

Often CSAs offer opportunities to visit, help out, and learn about growing practices and ecological relationships on farms (although members are certainly not required to do this, and many do not). Judith Belasco, who buys Fair Trade coffee for her household and has introduced Fair Trade tea and coffee to her office, sees Fair Trade and CSAs as parallel movements that can help each other grow. She connects CSAs and Fair Trade because both are about knowing the story behind one's food—where it comes from, who grows it, and what it takes to reach the consumer.[5] Pattie Cippi Harte, like Belasco, has brought Fair Trade tea and coffee to her urban workplace, and also speaks about the connections between CSAs and Fair Trade coffee. She thinks it is important to buy local because she gets better quality produce while reducing her carbon footprint.[6]

Cippi Harte emphasized the importance of buying locally because it reduces the environmental effects of our consumption choices in addition to bringing local consumer communities into closer contact with each other, with farmers, and with food.[7] These reinforced connections validate conscious consumers' beliefs and choices, which in turn strengthens their confidence in the power of consumers to effect positive changes in our food system. "Buy local" resonated with a number of the Fair Trade activists we spoke with. Some Fair Trade Towns encourage community members to buy Fair Trade products from locally owned businesses. "I don't say go to Wal-Mart to buy Fair Trade. I try to help people go to all these places, says Yuri Friman, pointing to locally owned businesses on a map.[8] "*Buy Local Buy Fair* is our slogan," says Alexandra Mello. "While you're supporting your local community, buy from fair trade sources."[9]

Both of the activists quoted above saw buying locally and buying fair as ways to strengthen the bonds within their own towns. Friman, for example, was proud to explain how his outreach efforts for Fair Trade Towns USA brought him into close working relationships with local newspapers, government, other nonprofit organizations and businesses—relationships that seemed to be mutually beneficial for the parties involved. He discussed his role in bringing farmers and consumers closer together through community talks with local farmers and travel programs in which participants visit Fair Trade farms overseas. Friman described himself as the "local face" because he organizes these talks and travel programs. When people see him at community events, they often recognize him as the "Fair Trade man." He enjoys the interaction with fellow community members while achieving his awareness-raising goals. He wants his community to "do well" monetarily and socially by supporting local farmers and retailers, and to "do good" by recognizing, respecting, and helping farmers who grow the products they must import, such as coffee.

Ideological Underpinnings

The ideological communities developing around Fair Trade in the United States are rooted in the joint power of consumer identities and social justice movements. Consumption has become a topic of social research because of its apparent role in identity formation (see, e.g., Featherstone 1991, Mansvelt 2007, Smith Maguire 2008, Soper 1999, Warde 1994). Whereas social class or family traditions have guided identity formation in the past, sociologist Ulrich Beck (1992) suggests that consumption practices now play a larger role in shaping ideas of self. At the same time, people have organized to try to change structures they see as unjust—in their own societies and in global trade relations (Buechler 2000). Fair Trade has created a way for people to identify as ethical consumers and to connect to others who are interested in reducing local and global inequalities.

Fair Trade Towns

Fair Trade Towns USA is a loosely tied network of communities around the United States that are committed to providing fairly traded products in their businesses and to raising awareness of the Fair Trade movement. There are currently twenty-four declared Fair Trade cities and towns across the United States (Fair Trade Towns USA 2011). To become a Fair Trade Town or Fair Trade City, organizers within a community must form a steering committee, reach out to area retailers to encourage them to carry Fair Trade products or help them expand their current product lines, engage the community (e.g., schools, hospitals, offices, and faith-based organizations), gain media attention, and pass a Fair Trade resolution in the town or city's council (or other governing body). The municipality resolves to support Fair Trade and the local campaign, and to choose Fair Trade products for meetings and events when there is a Fair Trade option (Fair Trade Towns USA 2010). The completion of these steps does not end the organizers' job; becoming recognized as a Fair Trade Town or Fair Trade City is just the beginning of an ongoing effort. Established Fair Trade towns or cities must document activity in at least one of Fair Trade Towns USA's list of potential "post-declaration campaigns" each year (items marked with an asterisk are considered by the national leadership to be "core campaigns"):

- Mentor a town through the declaration process*
- Participate in a Fair Trade Schools campaign*
- Participate in a Fair Trade Faith campaign*

* "Buy Fair, Buy Local" campaign involving local agriculture community
* Effectively organize and complete a trip to an origin of Fair Trade products
* Organize a regional event to pool resources with other Fair Trade Town communities
* Bring a representative from a Fair Trade artisan group, cooperative, or farm to your town
* Demonstrate an increase of Fair Trade products in local businesses with the goal of increasing the number of fully committed Fair Trade businesses in your locale
* Become an International Sister-Fair Trade Town/City
* Embark on a local or regional speaking tour about the benefits of achieving Fair Trade Town/City status
* Start a local or regional trade policy initiative campaign
* Raise your town or city's "Fair Trade Portfolio" by introducing new Fair Trade Certified products to the town as they become available (Fair Trade Towns USA 2010).

Some of the activists quoted here are active members of their own Fair Trade Town committee. They spoke about the benefits of working with other like-minded communities to achieve the mutual goal of promoting fairly traded products. In her efforts to convert her small city from one that has socially conscious consumers to a declared Fair Trade Town, with support from local government and a commitment to further the principles of Fair Trade, activist Alexandra Mello found a like-minded individual, Friman, in a community roughly half an hour away. They were able to work together to connect people in both communities around discussions about Fair Trade. With support from the national leadership, all of the declared Fair Trade Towns are working together to share their experiences and ideas. Friman senses that no town is left alone in its pursuit to expand the presence of fairly traded products and advance the ideology of ethical consumption, because regular

5.3 Fair Trade Towns Logo

dialogue allows activists to reflect on their individual and collective actions and identify each group's assets and on how to maximize them.

Statewide Fair Trade Coalitions

Statewide Fair Trade coalitions are another way to connect like-minded individuals. The Washington Fair Trade Coalition is made up of roughly forty-two businesses and organizations that work together to lobby the state and national government to change international trade policies so that they are more equitable and protective of human rights. Stephanie Celt, the coalition's former director, spoke about the particular importance of working together as groups. Each member represents an organization, and therefore consensus within the coalition requires understanding the dynamics of each member organization and its goals with respect to the coalition at large. Celt must understand the interpersonal relationships that are at play within the member organizations as well as those that emerge when member organizations come together. Similar to a cooperative, the coalition offers strength in numbers. Several organizations working together to achieve the same goal will prove to be more capable of effecting change.[10]

Celt comprehends the difficulties and also the benefits of organizing in this way. The Washington Fair Trade Coalition was founded after the "Battle of Seattle," when thousands of protesters from a broad range of organizations and institutions gathered to stop the 1999 World Trade Organization talks. This protest brought attention to unequal global trade relations and the social and environmental consequences of unfettered free trade, and engendered new coalitions (e.g., between Teamsters and environmental activists). The size and diversity of the Seattle protest highlighted the fact that transparent and ethical trade are not "fringe" issues but rather issues that people all over the world care deeply about. Similarly, the variety of organizations in the Washington Fair Trade Coalition, from government employees to an Audubon Society, represents diversity in civil society.

Faith

Faith is another ideological tie that brings people together to promote Fair Trade. Ethical traditions around food and human rights can inform consuming habits. Judith Belasco works for an organization called Hazon, whose mission statement is "creating a healthier and more sustainable Jewish community—as a step towards a healthier and more sustainable world for all." Hazon applies centuries-old Jewish traditions to develop and raise ethical awareness around food consumption. According to Jewish tradition, a field of food is to be used to provide nourishment, solace, and comfort to all those who may benefit from it. Hazon's work around food ethics strengthens interpersonal relationships through communities that believe in applying Jewish traditions to contemporary food issues, including the consumption of products that grow in the less-developed world. Hazon brings concern for food ethics to consumers who may not be thinking about the link between their consumption choices and their Judaism. This strategy creates stronger interpersonal relationships within the existing community of people who are connected by Jewish tradition and ecology. Hazon tightens the bonds of a community that already exists, bringing like-minded individuals together. Cippi Harte also works for an organization that employs Jewish traditions to build active communities and relationships within those communities. Like Belasco, she expresses clear links between Jewish laws and traditions and contemporary food ethics, particularly Fair Trade. She purchases Fair Trade products for her home and office because it reflects her value of respecting other people, which is rooted in Jewish law.

Faith provides an ideological link for consumer communities. The interviewees discussed here both happen to be Jewish, but the faith-fairness connection exists across many denominations and within groups of producers as well (see chapter 1). For example, Catholic Relief Services (CRS) trains "Fair Trade Ambassadors" to promote ethical consumption in their parishes and schools. Working with more than a dozen Fair Trade organizations, CRS ambassadors help Catholics to "shop with solidarity and build community" and to "think globally and act locally" by supporting Fair Trade. "Participating in CRS Fair Trade means supporting farmers and artisans, doing businesses with organizations you can believe in, and donating to our revolving Fund whose grants build a fair and sustainable marketplace" (Catholic Relief Services 2010).

Environmentalism

Environmental activists are keen to the fact that Fair Trade producer standards honor a nature-society link. Some of the Fair Trade activists emphasized the ecological and social benefits of community supported agriculture, which we have already discussed in the context of community building. Another example of an ecologically minded consumer community is cooperative markets that emphasize local seasonal produce, recycled goods, and nontoxic cleaning products. Ron Zisa, who purchases coffee and other bulk items for the United States' largest cooperative supermarket, the Park Slope Food Coop, in Brooklyn, New York, spoke about some of the common concerns at the ecologically minded cooperative. He recognizes that, while most members are concerned for their health and the health of those who grow their food, some members do not always see the direct links between the environment and their own purchases. Zisa aims to raise cooperative members' awareness of the environmental impact of food production and consumption and to help them make environmentally sustainable choices.[11]

Coffee roaster Barth Anderson's description of his trips to visit the farms from which he purchases coffee exemplify ecological consumption; he is most interested in people interacting with their unique ecosystems, working with particular plants, animals, and climatic conditions. It is the particular arrangements of these "ingredients" that make each of Anderson's coffees special to him and his customers. He is proud of the interpersonal relationships he develops with the farmers and with the people who buy coffee from him. He stresses human-animal-planet interactions; the specific ecosystems surrounding a given coffee plant contribute to what he purchases and then sells to his customers. Recognizing this importance of place, which includes the dynamic biodiversity that Anderson speaks of, is part of what makes Fair Trade a powerful tool for change.[12]

REDEFINING GLOBAL INTERPERSONAL RELATIONSHIPS

Since Fair Trade is a global project, many of the relationships formed around it are global as well. The success of ongoing Fair Trade relationships depends on "the formation and maintenance of networks of reciprocity, respect and trust" (Lyon 2007, 250). Co-op grocery manager Zisa's drawing illustrates the mutuality of an ideal Fair Trade relationship: "It works both ways. The buyer has responsibility to the farmers, and the farmers have a responsibility to the buyer also. It's a symbiotic relationship."[13]

Fair Trade is making it easier to maintain balanced relationships in trade because of the nature of the system and the personalities involved—people

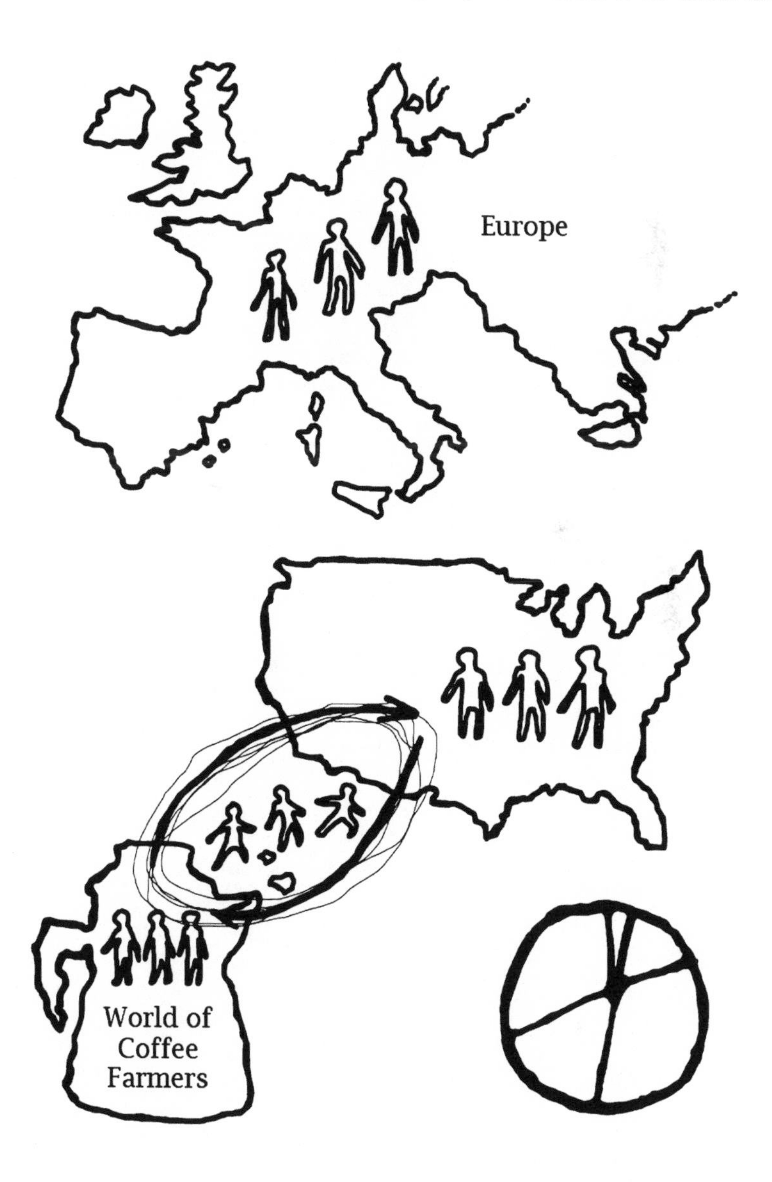

5.4 Drawing by Ron Zisa

who are genuinely interested in understanding their trading partners' lives and what they hope to gain from the trading relationships. These relationships go beyond the interests of trade. Alexandra Mello points to the desire to understand the lifestyles attached to coffee growing and coffee consump-

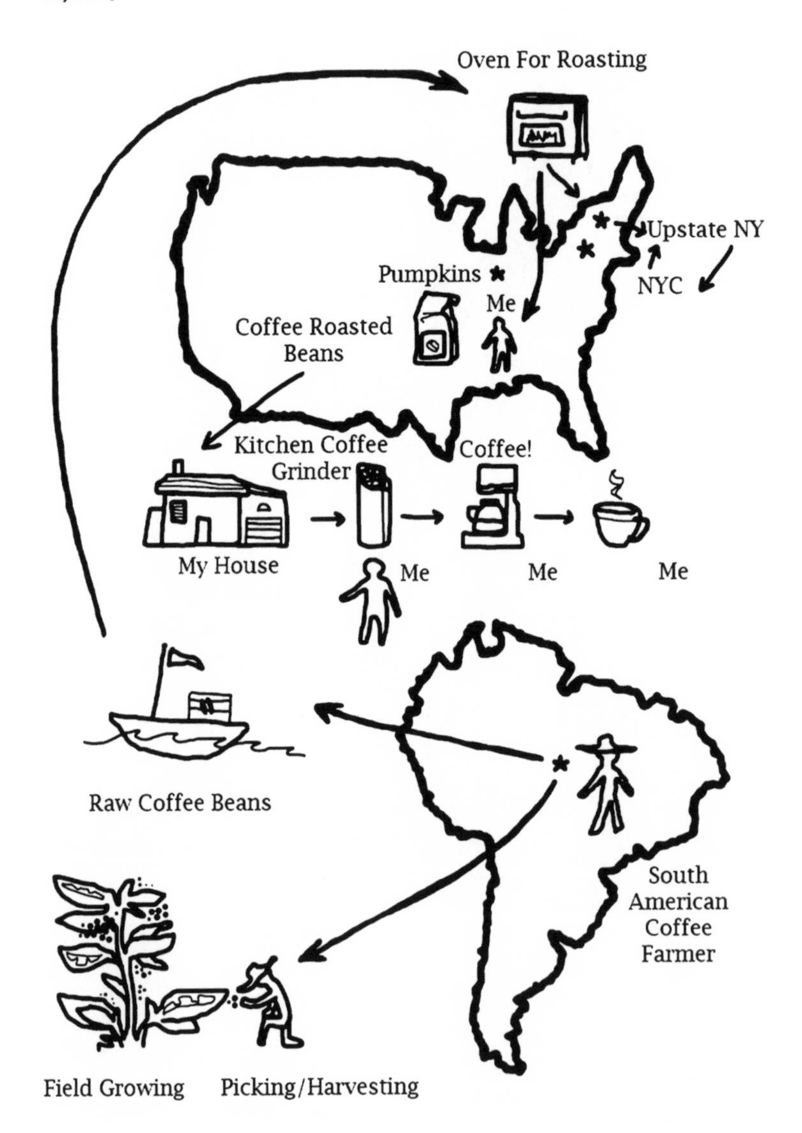

5.5 Drawing by Alexandra Mello

tion. She would like to see mutual understanding between people in coffee-growing and coffee-consuming regions.

Mello illustrates the details and benefits of taking trading relationships personally, taking the time to recognize the people on the producing end of the value chain. She posits that such change is necessary and possible with

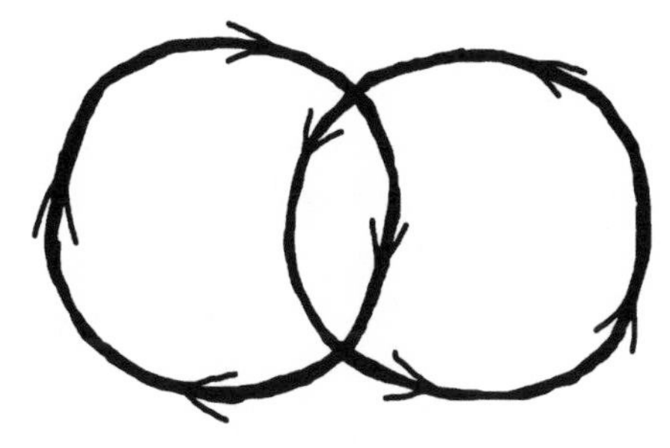

5.6 Drawing by Judith Belasco

the Fair Trade model. It is a start toward reaching a meaningful understanding of people and cultures outside of our own. She emphasizes that more work is needed to further the connections and bridges that activists, traders, businesses, farmers, and consumers are building. The sustainability and human rights aspects of Fair Trade will persist and progress only to the degree that local and global community bonds are continually negotiated.

Belasco drew herself as part of the Fair Trade supply chain. Like Cippi Harte, she keeps herself in her urban environment but draws lines connecting her and her home to the store where she purchases Fair Trade coffee, and also to the roaster, the port, and the origin of the coffee.

Anderson, who has firsthand knowledge of the communities and farms he buys from, exemplifies how a deeper local-global relationship can play out in practice. He visits coffee farmers to deepen his own understanding of the business he is in. He is quick to mention that these are not formal quality-assurance trips; he is going not as a boss but as an interested colleague. Anderson is also extremely grounded in the coffee-consumer community where he lives and works. He offers workshops to those who will sell the coffee that he imports and roasts so that they can learn to appreciate various aspects of the product that speak to the growing community's culture. While he may not be typical of a Fair Trade activist or coffee roaster, Anderson illustrates that it is possible for one person to maintain strong global *and* local relationships.

Implications for Development

Fair Trade networks and value chains promote global citizenship in that participants begin to see their obligations to other people in a way that is not influenced by physical distance or national borders. This opens the way for producers and workers to have a powerful voice in their businesses and in the future of their communities. With empowerment comes a sense of value, worth, and capability (Sen 1999). The presence of more voices in global value chains also means a mitigation of homogenizing cultural norms and expectations (DeCarlo 2007).

Since the establishment of the World Bank and the International Monetary Fund after World War II, the field of development has focused on economic growth in order to bring countries in line with the world's wealthiest nations—economically and, subsequently, culturally. Fair Trade is an example of resistance to this notion of development in that it has brought noneconomic factors such as human rights, environmental protection, and local cultural contexts to the fore. It is market-based but popularizes nonhegemonic models and ideals. Within Fair Trade there is no single definition of development or efficiency or economic production (McGregor 2007). This is a practical application of what scholars call "post-development" theory, which posits that development is an unyielding process of change (McKinnon 2007) and contests static definitions of terms such as "development," "human rights," and "community" (Feagan 2007, Santos 2003). Fair Trade activists are applying post-development theory in that they are fighting for the localization of development discourse. They are rejecting the unequal trade norms that they were taught to accept as regular or correct and are instead promoting the role of farmers in the global South as agents of change.

Discussions Fuel a Cycle of Change

Maybe the most important break from hegemony has been, and will continue to be, the conversations taking place within activist and consumer communities. Within these physical and ideological communities, people must keep on openly engaging with each other and with their trading partners. Members of a group, such as a Fair Trade network, must decide on rules for mutual respect and understanding. Constant dialogue is necessary to achieve consensus because the members of the network are also members of other networks (be they diverse cultural, religious, country, regional, or other groups), each with their own sets of standards. The borders of each network are constantly being transgressed. What is each person willing to sacrifice to ensure that all members have their basic needs met? Members must come to an agreement on this in order for the group or network to succeed. When a value chain consists of individuals in various locations and countries, and when some individuals in this value chain are not getting their basic needs met, as has been the case with poverty-and debt-stricken farmers around the world, it becomes the obligation of group members to discuss a redistribution of duties so that at least all basic needs are met.

Farmers, importers, retailers, and end consumers all must reach outside their local communities and recognize the shared relationship they have with each other if they are all going to enjoy the benefits of that relationship. Membership in a group means security, self-esteem, and identity but also the responsibility of mutual respect in order to maintain cohesion (Brock 2005). The ethical responsibilities of protecting the sanctity of these relationships do not diminish with distance. It is not only possible but almost mandatory that we take a broader view of our obligations to other people. Fair Trade seeks to actualize this inevitability. Local outreach programs, efforts to physically connect producers and consumers, political activism, and the *willingness to engage in continued dialogue* all fuel a positive cycle of change.

Discussions within and between global-North and global-South communities must continue to serve as evaluative tools for reflection because they increase the probability of nonhegemonic decision making. Each community focuses on its social, cultural, and ecological assets and on how to make the most of them economically without damaging an asset's source or exploiting those of another community. What are we doing right, and how can we strengthen this? What do we need to improve? What should we change?

5.7 Drawing by Pattie Cippi Harte

Cippi Harte recognizes that relationships within Fair Trade networks are dynamic. She points out that accepting the ever-changing nature of relationships within Fair Trade value chains means that there is never one particular endpoint that all stakeholders aspire to reach. The circles and arrows in her drawing illustrate how important it is to focus on principals, such as creating a system of trade that allows farmers to live the lives they choose, and then to frequently reevaluate how best to put these principles into action.

CONCLUSION

The future of the Fair Trade movement depends on continual interaction and reflexive discussion (DuPuis and Gillon [2009] state the same in regard to the organics movement). It is not the hope that all the communities involved in Fair Trade will grow so large that they eventually become one global ideological community. The importance of place and context prevail because global phenomena are localized in practice (Escobar 2001, McGregor 2007). Communities of Fair Trade supporters in the United States and elsewhere can use their unique positions as business people and educated, concerned, and ethical shoppers to maximize the power of Fair Trade consumption in terms of its impacts in the developing world. Reflexively discussing the impacts of relationships, whether they be trading relationships or common consuming relationships, offers the space for recognizing the power and the ability to effect change. Maintaining open, transparent conversations about consumption can bring notions of human rights into discussions where they have previously been left unmentioned. The more people engage in conversations about the effects, implications, and power of their consumption habits, the more momentum will be available to fuel a positive cycle of change, bringing more awareness of the need for equality and respect for human rights in trade.

Fair Trade activists are promoting community in varied ideological circles, notable Fair Trade Towns, faith-based organizations, environmental groups, and larger coalitions of like-minded individuals. Interaction within and among these groups fuels dialogue about redefining global interpersonal relations in ways that pay attention to human rights and environmental sustainability, and recognize interdependence at the global and local levels. This chapter brings some of the important consequences of Fair

Trade relationships to the fore. In the pursuit of more ethical, transparent, and sustainable ways to consume, local communities are being strengthened. At the same time, global relationships are changed and challenged as value chains are transformed from narrow business relationships with low transparency and high power asymmetry to sustainable relationships built on respect and fairness.

Chapter 6

A Fair Trade University

In spring 2010, my workplace, the University of California at San Diego (UCSD), became the second Fair Trade University in the United States.[1] All coffee and tea, most sugar, and some of the rice and quinoa served in our university-managed facilities is Fair Trade Certified. Venues that sell chocolate (including beverages and ice cream) offer at least one Fair Trade option. The Fair Trade label is prominently displayed wherever these products are served or sold. Student activists provide training to the employees of on-campus vendors about the benefits of Fair Trade. Perhaps most importantly, the agreement between the students and the university employees responsible for purchasing, housing and dining operations, and lease negotiations with privately owned food vendors stipulates a continued collaborative effort to increase the availability of Fair Trade Certified products on the campus and to promote social and environmental sustainability in other ways, such as sourcing from local organic farmers and reducing waste.

When I arrived on campus in 2004, recyclable plastic water bottles bobbed in the trash cans. It was nearly impossible to find Fair Trade coffee (let alone other products), and challenging enough to buy a meal served on a nondisposable plate. Today UCSD is committed to environmental sustainability and ethical trade. This chapter is a case study of a successful, student-initiated campaign that has made a big difference, and of how UCSD students and administrators have made promoting "a better quality of life for people they probably will never meet"[2] part of their core mission.

I first provide a bit of background on university-based Fair Trade campaigns and then chronicle the UCSD effort in some detail. The reflections and comments constitute a story of innovation, persistence, connection, and vision. Indeed, this chapter can be read as a blueprint as well as a story.[3]

FAIR TRADE GOES TO SCHOOL

UCSD is certainly not the first school where students have asked for, and gotten, Fair Trade products in campus food-service establishments. United Students for Fair Trade (USFT) has been helping UK and US students to do this since 1995. In the beginning, students' efforts to promote Fair Trade were piecemeal, and in the United States they remained so until quite recently. Sometimes organizers sought only to raise awareness, for example, by sponsoring a Fair Trade farmer to speak at their school or by throwing a "Fair Trade fest." But others have effected changes in university-managed food service operations. For example,

> In a tremendous victory for campus activists, Brandeis University became the latest in a growing number of colleges and universities to switch some of its food establishments to solely Fair Trade Certified coffee. Two dining halls and a campus convenience store now exclusively offer Fair Trade Certified coffee. This change comes on the heels of the overwhelming passage of an undergraduate student body referendum on switching to solely Fair Trade coffee.[4]

Other student campaigns have persuaded privately operated vendors to offer Fair Trade products (usually coffee). For instance, at the University of Florida at Gainesville, "Students making trade fair held a Fair Trade fair featuring a giant coffee mug, Fair Trade Certified food and information, magic shows and entertainment."[5] The organization gathered at least 200 signatures on a petition to encourage an on-campus café to offer Fair Trade espresso.

The concept of an integrated "Fairtrade School" emerged in the United Kingdom, where the Fairtrade Foundation now maintains a registry of schools, colleges, and universities that:

1 Have set up a Fairtrade School Steering Group.
2 Have written and adopted a whole-school Fairtrade Policy.
3 Are committed to selling, promoting and using Fairtrade products.
4 Learn about Fairtrade issues.
5 Promote and take action for Fairtrade both in school and the wider community (Fairtrade Foundation 2009c).

In 2008 students and administrators at the University of Wisconsin Oshkosh followed this blueprint and declared themselves a Fair Trade University, although as a US school they could not be listed in the Fairtrade Foundation's registry. Student activists at UCSD saw the Fairtrade Foundation goals as good guidelines, but as their efforts matured, they realized that they wanted to (1) be more specific about the scope of their resolution so that it would in fact change the way the university and companies with on-campus franchises do business, and (2) mandate continual evaluation and improvement. Importantly, the second goal concurs with Fairtrade International's guidelines for producers.

Box 6.1

UC San Diego Fair Trade University Policy, 2010

I POLICY

In addition to the UC Presidential Policy on Sustainable Practices, UC San Diego, in 2009, made the decision to promote social responsibility by demonstrating a commitment to the principles of Fair Trade. This is achieved through working towards a goal of exclusive offerings of selected Fair Trade Certified™ products throughout the main campus. By making a commitment to promote Fair Trade Certified™ products, UC San Diego supports livable wages, humane working conditions, and guarantees against the use of child labor. In cooperation with TransFair, UC San Diego is pursuing the status of a "Fair Trade University."

II PRACTICES

In support of this policy, UC San Diego adopts the following practices:

A Exclusively Fair Trade Certified™ Products

At every self-operated campus foodservice operation location where the following products are sold or served, these products must be 100% Fair Trade Certified™:

1 Coffees, including caffeinated, decaffeinated, and espressos, served brewed hot or iced or sold in packages.
2 Teas, including caffeinated, decaffeinated, brewed hot or iced or sold bagged and loose packaged, but excluding teas sold prepared and prepackaged in bottles or from vending machines or beverage stations.
3 Sugars, including granular, raw cane, brown sugar, powdered sugar, Demerara, Molasses, and Sucanat, sold or served in bulk or in individual packets.

UC San Diego agrees to work toward the goal of implementing this policy with respect to contracted foodservice operations on the main campus upon contract renewals and requests for proposals for foodservice operators.

B Fair Trade Certified™ Product Options

At every self-operated campus foodservice operation location where the following products are sold or served, at least one Fair Trade Certified™ option must be available for consumers:

1 Chocolate, including chocolate candy bars, hot chocolate, and mocha drinks.
2 Ice Cream Pints: At least one variety made with Fair Trade Certified™ ingredients. If provider offers flavor in two versions, one certified, vendor will only offer certified version.
3 Grains: Rice: including (but not limited to) White, Brown, Thai, Coral, Ruby, or Purple;
4 Quinoa: White, Red, Black

UC San Diego agrees to work toward the goal of implementing this policy with respect to contracted foodservice operations on the main campus upon contract renewals and requests for proposals for contracted foodservice operators.

C Information and Advertising

1 Official Fair Trade Certified™ label shall be displayed in the immediate and visible vicinity where products are served and/or sold.
2 Signage shall be displayed by main entrance or cashier informing customers that the foodservice operation offers Fair Trade products.

D Employee Training

1 Student representatives from the Fair Trade Advisory Committee will help with initial training and develop program for future trainings in order to sustain continued education.

2 All employees of applicable campus foodservice operations will be provided training materials on the benefits of Fair Trade Certification™ and availability of the Fair Trade products provided by that particular operation.

E Continued Collaborative Effort: The Fair Trade Advisory Committee will guarantee student input during discussions regarding changes or additions to the Fair Trade Policy as it relates to Fair Trade food and beverage contracts and Requests for Proposals. The Chair of the Fair Trade Advisory Committee will ensure that this policy is provided to purchasing and contracting to be included in all appropriate Requests for Proposals and contracts.

F Fair Trade Advisory Committee

1 Membership

Committee shall include:

- 2 undergraduate students appointed by the Student Sustainability Collective
- 1 undergraduate student appointed by the Associated Students Vice President of External Affairs
- 1 student at large representative appointed by All Campus Commuter Board (ACCB)
- 1 student who resides on campus appointed by Inter College Residents Association (ICRA)
- 1 graduate student appointed by the Graduate Student Association
- 1 student representative appointed by the Co-Op Union
- Director or appointed permanent staff member from the Cross Cultural Center
- A University Centers representative appointed by the Director of University Centers
- A representative from Housing, Dining, and Hospitality
- A representative from the Bookstore
- A representative from Real Estate

If an unrepresented entity applies for a position on the Fair Trade Advisory Committee, said committee can create a new seat to represent the applying entity with a two-thirds majority vote.

2 Duties and Powers

a. The committee shall meet at least twice a quarter.

b. The Advisory Committee will ensure the University Fair Trade Policy is upheld and advanced.

c. The Advisory Committee shall advise and make recommendations to the University on issues related to the Fair Trade Policy, including its administration.

d. The Advisory Committee will conduct research into new Fair Trade products.

  1) This research work shall be carried out by the Fair Trade intern and a Fair Trade Policy subcommittee, whose membership will have a student majority and will be appointed by the Advisory Committee.

  2) The intern shall be unpaid, and appointed by a subcommittee of the Fair Trade Advisory Committee student representatives and the Fair Trade Advisory Committee Chair.

  3) The intern's responsibilities include:

    a) Chairing the Fair Trade Policy subcommittee,

    b) Updating the Fair Trade website with findings of the Policy subcommittee,

    c) Collaborating with the Fair Trade Advisory Committee Chair,

    d) Working with the Fair Trade Advisory Committee student representatives of the Student Sustainability Collective to ensure that the UC San Diego Fair Trade Policy is upheld and advanced.

e. The Advisory Committee shall post an annual report on the fairtrade.ucsd.edu website stating the status of Fair Trade at UC San Diego. This report will also be sent to:

  1) The Chancellor's Office

  2) Vice Chancellor, External & Business Affairs

  3) Vice Chancellor, Student Affairs

  4) The Associated Students

  5) The Graduate Student Association

  6) UCSD Real Estate

  7) TransFair USA

f. Educational Materials, at a minimum:

  1) "Take one" pamphlets on the benefits Fair Trade as offered by TransFair,

  2) Posters if requested by foodservice operators,

  3) UC San Diego Fair Trade website (fairtrade.ucsd.edu).

g. Collaboration

  1) The Advisory Committee shall provide recognition to foodservice opera-

tors who have reached levels of compliance with respect to the Fair Trade Policy, as determined by the Advisory Committee.

2) The Advisory Committee shall be responsible for maintaining a positive, collaborative relationship with foodservice operators in order to share questions and information between the two groups.

G Growth

1 Contracts

UC San Diego agrees to work toward the goal of implementing this policy with respect to contracted foodservice operations on the main campus upon contract renewals and requests for proposals from foodservice operators.

2 New Products

a. The Fair Trade Advisory Committee will determine if a Fair Trade Certified product can feasibly replace a non-certified product.

b. As new Fair Trade Certified products become available on the wider market, they shall be made available as options at foodservice operations on the main campus until such a time as they can replace the non-certified items.

F Initiation

Upon the implementation date of the Fair Trade University Policy, all new contracts to be negotiated or to be renewed after that date for self-operated foodservice operations and contract foodservice operations will meet or exceed the above criteria. This policy can be changed with a two-thirds majority vote of the Fair Trade Advisory Committee members at two consecutive meetings.

Fair Trade at UCSD: A Timeline

In 2003, at UCSD, One Earth One Justice (OEOJ) was born. OEOJ was founded as an active social justice group that raises awareness of environmental and social injustices and seeks to empower people with opportunities to create change. Promoting Fair Trade was central to OEOJ's mission because, as one member put it, Fair Trade "gets to the heart of a lot of issues about self-empowerment and sustainability and economic justice."[6]

Much of the initial energy behind establishing OEOJ and launching a Fair Trade campaign came from five individuals: Jeremy Linneman, a

6.1 Logo for One Earth One Justice T-shirt

graduate student in international studies; Viraf Soroushian, an undergraduate majoring in sociology and economics; his roommate, political science major Jeremy Seymour; and their classmates Kate Maull and Boshen Jia. Linneman began working as a teaching assistant for a required yearlong freshman writing course concerned with social justice, the history and cultural experience of underrepresented groups, and the development of intelligent citizenship—topics he had not thought much about as an undergraduate. Soroushian and Maull were his students.

Linneman coined OEOJ's guiding theme "Think, Learn, Change, Grow." While he was establishing the group as a nonprofit organization,[7] the four undergraduates, led by Soroushian, tried to solidify its existence at UCSD. Linneman recalled his early relationship with his student and friend Soroushian:

> For me, it started with, "This is economic democracy; we're putting forth this idea that how people spend their dollars makes a difference." And Viraf would respond, "How can anyone spend their dollars differently at UCSD? There are no choices here. You're telling these people they need to vote with their dollars, but there are no options for them." And so I told him, "All right, then, you do something about it, man," and he did. He went out and he got a bunch of people together and he came up with the whole "party with a purpose" idea and got everyone together to start talking about Fair Trade coffee. Fair Trade coffee and Fair Trade bananas were his two original ideas. And Viraf ran with it, and he pushed it and he pushed it and he won.

Here is Soroushian's piece of the story:

> [Linneman] was such a great TA. He had this uncanny way of relating to his students and challenging their minds. He created the concept of One Earth One Justice. It's to educate youth about the connection between their lives and the social and environmental crises happening in the world around them. The first campaign that we decided to run was something we called "economic democracy." It's the idea of using your money as a tool to make change in the world, so to use every dollar—your everyday spending—as a vote for a better world. This concept could have taken us in many different directions, but the direction that we felt stuck the best on a college campus was through Fair Trade. It was something that I had heard about, but another friend, a fellow founder of OEOJ, Jeremy Seymour, was the one who put forth the idea. When we started our campaign, there were only two dining halls carrying Fair Trade coffee, and I think maybe one [student-operated campus café] was also carrying it. . . . We were young, inexperienced, and we had no idea about how to do anything or how to make change happen.

Seymour added,

> When we first started out, our mission was to look for an issue that focused on interconnecting environmental and social justice problems that we could do something about on campus and get people involved in. Fair Trade really crossed the borders as something that all of us across the board were very, very interested in. We spent a lot of time out tabling, doing coffee stands, and then we started to spread our group, hoping to be a larger umbrella organization that could reach out to other groups to form alliances on the campus.

Meanwhile, Maull—OEOJ's comptroller—involved herself in student government and introduced a Fair Trade resolution to the Associated Students (A.S.). She remembers that

> when the very first Fair Trade resolution passed in A.S., the main issue was "Why should A.S. make a statement about this?" And so people like Viraf and I said, "We have a stake as business owners."[8] We buy prod-

> ucts to sell to students, and so it's in our interest to make students aware that they have the option to purchase things that have a lower impact on the environment or a better socioeconomic impact in the world, and we should be giving people an educated choice.

By fall 2005 there was a Fair Trade coffee option in each of UCSD's five dining halls, on a trial basis. Undergraduate and graduate student governing bodies had passed nonbinding resolutions to encourage the university to purchase more Fair Trade products. Soroushian and Seymour gave presentations at campus events and were invited to speak in the writing course where the idea began. But despite OEOJ's successes, UCSD administrators were lukewarm about moving forward with Fair Trade. According to Seymour,

> The administration's position when we first approached them about Fair Trade was, "Look, students don't care. All our students care about really is the price." So of course our initial response was, "Well, *we're students*." We got together and planned ways to show them that students care. That's where a lot of the tabling came in, and we did some petitions; we had lots of students' names saying "I want Fair Trade coffee." Once the Fair Trade was there, we followed up and let students know what Fair Trade coffee was—what the label was. All the coffee shops had the TransFair USA labels out so people could know what it was and have that visual recognition. We continued to promote it and make sure that it stayed there, because it was originally just a trial. When we had really kicked this thing into high gear, Viraf and I were juniors. So we had two years.

Soroushian added,

> We were also splitting our time with establishing the roots of OEOJ and UCSD's activist community, planning the "party with a purpose" events, and training and recruiting the younger generation. Other groups, like Students for Economic Justice, which dedicated all their energy to the SweatFree campaign, succeeded in much less time but became defunct after the core members graduated. We wanted activism at UCSD to grow and build from what we started.

In winter 2006, OEOJ members presented a letter bearing more than five hundred signatures to Chancellor Marye Anne Fox. It asked that all coffee vendors at UCSD carry at least one Fair Trade Certified option. This was not the first time that Fox, who had come to UCSD in 2004, had heard from OEOJ. During her first year, she scheduled regular town hall meetings with students. According to Soroushian,

> OEOJ had been trying to arrange a meeting with [Fox] for about a quarter and a half, with very little success. The week before the town hall meeting, she had given us a date, but it was the Friday before finals week. The end of the quarter is a very stressful time for the student activist; we were trying to tie up all our advocacy work so that we could focus on our studies. Meetings like this require a lot of prep work and a time afterwards to stew over and follow up on what was discussed. And you lose all that momentum during finals week and the break. So . . . we saw the town hall meeting as an appropriate and timely forum to make our case to her. We also used this as an opportunity to show the unity of the activist communities, and we mobilized ally organizations to participate in the meeting. We all wore a green armband as a sign of solidarity. It turned out that the forty to forty-five of us that attended made up about 90 percent of the meeting! It was great. Boshen [Jia] and I were to make a prepared statement on behalf of the group. I went up first. And when I said, "The following students believe UCSD can do more to be socially and environmentally sustainable," everybody stood up. I asked what the administration was doing to address this concern. Chancellor Fox opened the floor to her senior staff, who were also present, and they went over a number of initiatives they were doing, but most of them focused on environmental sustainability, which was a concern of ours going in. It's not uncommon for institutions and corporations to use their environmental sustainability agenda to distract attention from, or greenwash, social-justice and labor-rights issues. So, we made it clear to her that social justice, particularly Fair Trade, was a valid concern. Then Boshen went up and demanded that she take our request to meet with her seriously. What came out of that action was that we and three other organizations sat down with the chancellor for about an hour and each told her about our campaigns and specifically what we were asking her and her staff to do. She passed on our message to senior staff members. After that, the Sunshine Store in the Price Center and the cart by Mandeville auditorium started carrying Fair Trade coffee.

According to Fox,

> The thing that was most impressive was their passion. They thought this was something that was going to change the world, and perhaps it will. . . . They came in as a group and sat down on the couches and made themselves at home. They had brought a big sheet cake. I said, "You know, guys, you're supposed to ask me for something and then I say no, and then you come and bring me the cake. You skipped the first step." It was a good conversation, and I think we've worked well together since. They gave me a tutorial on the whole field of Fair Trade and where we were on each of our facilities that used the products, and it was enlightening to me because I hadn't looked at it in that much detail.

The following academic year, OEOJ members convinced the University Centers Advisory Board to preference Fair Trade vendors in the bidding process to fill a new coffee shop space. The group worked with the UCSD bookstore to establish the first 100 percent Fair Trade coffee vendor on campus (see box 6.2). One Earth One Justice member June Reyes, who had recently joined the group, remembers thinking, "Wow—we can really do anything we want on campus!"

To link with other activists, raise students' awareness, and build support for Fair Trade at UCSD, OEOJ organized another "party with a purpose," set up a Fair Trade Café on the campus's busiest pedestrian thoroughfare, arranged a screening of the documentary *Black Gold* (about how the Fair Trade movement has changed rural communities in Ethiopia), and sponsored a multifaceted celebration of Fair Trade month (in October). But perhaps OEOJ's most popular outreach activity was the ice cream giveaway. Member Rishi Ghosh told me,

> We do events where we'll raise awareness about Fair Trade with 300, 400 people for a day. We get Ben & Jerry's Fair Trade ice cream and give it out for free on campus. This is a decision we made a long time back, that we're not going to sell the product, we're going to give it away for free and just have a petition there for people to sign, and we'll absorb the loss. When we give out Fair Trade ice cream we make them stand in line on purpose, and we essentially preach Fair Trade for a few seconds, in a nice way. We show them the logos, and we tell them what to look for in the

> bookstore and in the coffee places and the dining halls. I get a feeling that they're there for the ice cream, and sometimes it's just not effective . . . but we try, and we do have our moments. If we've had a meeting with the administration the week before, we've told them there's going to be free Fair Trade ice cream on Library Walk this coming Friday, and we'll give you two scoops each. Having these people drop by at our events shows them that we are pursuing this relentlessly and that we're not just some student org that's going to kind of die off so they can delay us until we forget about what we are doing. It involves them a bit, it makes them feel a bit special to get the free ice cream.

By spring 2008, OEOJ had collected over 1,000 signatures in support of expanding the Fair Trade options on campus. That summer, the university's Housing, Dining, and Hospitality (HDH) unit, which oversees the dining halls, announced that it would start using Peet's Fair Trade Blend for all brewed coffee. Behind this move was HDH director Mark Cunningham, who has become a champion for social and environmental sustainability on campus. With resident students' approval, in August 2008 Cunningham was able to hire Krista Mays, who had been running UCSD's recycling, waste-reduction, energy-conservation, and water-conservation projects, as HDH's first sustainability manager.

At first Mays focused on the environmental aspects of HDH. She recalled that,

> As the students started to roll social responsibility and social sustainability more into the whole sustainability movement, that got moved in as well. Originally, I don't think I was thinking that sustainability would include Fair Trade and cage-free eggs, but it definitely does. . . . [The OEOJ students] do a pretty good job of getting the information and gathering the beginnings of critical mass and then bringing it forward. They're definitely more organized now than they used to be on these kinds of issues. They can get a group together faster, and they're more connected.

By this time UCSD had launched an Environment and Sustainability Initiative (including seminars to help professors incorporate sustainability into their courses and community outreach events) and declared the goal of becoming the greenest campus in the world. An advisory committee on

Box 6.2

Good Coffee at the Bookstore

JOHN TURK, UCSD BOOKSTORE DIRECTOR

We had an opportunity to build a coffee shop, and none of us knew how to run a coffee shop. We had a lot of help. We had volunteers coming out of the woodwork on campus that said, "Hire me! Hire me!" but we took a proactive approach to it and we did a taste test. We found four vendors that were willing to partner with us to provide product, R&D, a pumpkin latte in the fall, and trend drinks like smoothies. We needed good taste and rigorous and frequent training for our staff, because we wanted to run it. And we wanted to be able to employ students in this operation as well as our career staff. We wanted the service quality in the store to be the same in the coffee shop.

We didn't know if the one Fair Trade vendor would taste better or equal. We didn't have any idea how this was all going to turn out. But after blind tasting with sixty people over four days: light, medium, bold roast, caffeinated, decaf, tea . . . Barefoot Roasters [the Fair Trade company] was the number one in taste. We were happy about that, and they were thrilled.

So we've been learning the coffee shop business two steps forward1.8 steps backwards. It's quite different than the retail that we're used to. It's been successful and continues to grow. And I think it has a pretty good reputation on campus for comfort and the ambiance and being Fair Trade. And a lot of people like the way the drinks are made. They're not made from formulas or premix. It's all from scratch.

Interview by A. Linton, June 5, 2009.

sustainability was in place; it established six principles of sustainability to guide the university's progress, including recognition that sustainability is a key part of the school's mission, and that sustainability encompasses social, economic, and cultural interactions set within a supporting ecosystem. The committee also recommended—and got—a new permanent position: a UCSD sustainability coordinator.

Chancellor Fox spoke of the evolution of UCSD's sustainability mission as

> a way in which we could bring different parts of the faculty and the administration together. . . . When I came, I said that I wanted to pursue three I's: *interdisciplinary* work, which is certainly involved in solar energy conversion; *innovative* work, which means commercialization and finding ways that an application can go forward; and *international* collaborations. It's been possible to do all those by having it, this sustainability, as a central theme.

In fall 2008, OEOJ members and Associated Students president Donna Bean met with Chancellor Fox to discuss making UCSD a Fair Trade University. Without hesitation, the chancellor established an advisory committee composed of members of OEOJ and representatives from HDH, University Centers, UCSD Real Estate, and the campus bookstore.[9] The committee, chaired by Mark Cunningham, began meeting on a weekly basis in winter 2009. By the end of the academic year, they had laid the groundwork for Fair Trade awareness across campus by meeting with on-campus vendors not connected to HDH and drafting a charter for the Fair Trade University Advisory Board.

Sustainability 2.0

My interviews with UCSD administrators and staff invariably veered "off course" into discussions about composting, solar energy, water conservation, bags at the bookstore, giving every freshman a sturdy water bottle, reusable food containers, recycling, cage-free eggs, sourcing local and organic foods, and waste management. . . . I soon realized that these conversations were not digressions at all, because my interviewees—all of whom interfaced with the Fair Trade Advisory Board—saw Fair Trade as part of a much bigger picture that is coming into focus all around them. This is at least in part because, in January 2009, Chancellor Fox announced the establishment of UCSD's Sustainability 2.0 Initiative:

> This initiative builds on our historically strong disciplinary and interdisciplinary research enterprise and the foundation established by our campus that has helped unravel the causes of global warming. UC San Diego is now creating an umbrella for a coordinated campus-wide effort to advance

> our understanding of the scientific, economic, and social issues involved in the sustainable stewardship of planet earth in the twenty-first century.

Asked if they thought Sustainability 2.0 influenced the administration's willingness to commit to sourcing Fair Trade products at UCSD after years of steady but very slow progress, OEOJ students told me,

> I think it influenced it a lot, because the administration wasn't as willing to do Fair Trade, or listen to Fair Trade, until the connection was made between Fair Trade and sustainability and the environment, and what the chancellor set as the university's goal. And now that serves as a kind of stepping-stone for us to have these stakeholders come together and say, yes, Fair Trade is sustainability.[10]
>
> Fair Trade gets kind of absorbed into the whole "let's be green" movement. But it is very powerful at the same time.[11]
>
> It's got to help—we use that to hold them accountable. . . . But the pivotal point in getting the administration to switch their mindset was when we ran to the chancellor. Once we did that, the Fair Trade Committee was formed almost within two weeks. And with Mark Cunningham, the head of Housing and Dining, on board, he gives the directives to his employees and it has become much easier for us.[12]
>
> We really had our messaging points down and capitalized on them. We owed a lot to the environmental movement nationwide, and also to past advocacy about Fair Trade, the fact that it was already in the dining halls. *But with the sustainability aspect, we could take that and then add on the aspect of social sustainability.*[13]

Regarding her directive to make establishing UCSD as a Fair Trade University part of various administrators' jobs, Chancellor Fox emphasized that "by then [fall 2008], we had sustainability as an umbrella." Plus, "the students in the last year or two have been much more persistent than the ones earlier. They're more politically sophisticated. And, of course, there's very strong leadership in Housing and Dining."

Cunningham, the source of that leadership, added,

> "Sustainable" means that in five years we're still doing it. A university is kind of like a glacier. Once you get it set, it just creeps along. But it also recedes a lot, and you can fall back quickly. Particularly in hard times, you can say, "We can't afford that; we can't afford that; we can't afford that." To be a better place and a better program, you really do need to think globally, and have a champion in the administration.

SUCCESSFUL STRATEGIES

University of California San Diego and One Earth One Justice's "blueprint" for a successful Fair Trade campaign is composed of several elements: demonstrated student support; the proposal of concrete, doable alternatives to the status quo; a presence within student government and with the administration; and reasonableness, patience, and persistence. Effective strategies require proactive effort.

Collect Evidence

University administrators and on-campus businesses will be more willing to enact a proposed change when there is evidence of student demand. A petition asking for more Fair Trade products on campus was always on hand at OEOJ's public events. The group also surveyed over 400 students, asking about waste reduction and sustainability, including whether respondents were willing to pay more for Fair Trade coffee in the dining halls. They presented the results to Dining Management. According to Yvonne Macon, HDH's senior manager for procurement and contracts, the survey "was timely and very helpful in our effort to reduce disposable use. . . . We were already serving sustainably procured coffee, and some chocolates, etc., at that time, so that information was useful as we moved forward. It was very well done."[14] Sustainability is now included as a "quality point" assigned to products when HDH makes cost comparisons.

Do the Work

A successful institutional campaign involves providing purchasing agents with viable sourcing alternatives, for example, suppliers of Fair Trade products. One Earth One Justice members were quick to do this, and to pick up other work the advisory committee generated, such as taking and distributing meeting minutes.

> We've been very, very willing from the beginning to essentially do the grunt work. We avoid the stereotypes a lot of people have about activists or progressive groups. We're on top of our e-mail; we're professional in our communications; we're not afraid to pick up the phone and have a conversation; we try to keep our website organized.[15]

Build Bridges

> I'm the kind of person who likes to work from within. I like to negotiate with those who are in power in order to get better results, get more access to the system. So I think what we did right was that we took two approaches at the same time. We really reached out to school administrators *and* to the student body.[16]

One Earth One Justice members produced and distributed high-quality informational pamphlets about Fair Trade, but they also realized that most students are not in the mood to do more reading. Early on it became clear to OEOJ members that they needed to maximize the group's exposure in settings where they could actually talk with people, not just hand out fliers. They did this by participating in student government and by joining the student advisory committees of the university's six residential colleges and of the University Center.

Meanwhile, some lessons were learned about reaching administrators who could make changes happen:

> We've found that escalating [the Fair Trade University proposal] to higher and higher authority figures faster saves us months and months of having meetings with the purchasing director, and helping her through every single purchasing paper, and yet coming to a situation where we need to talk to someone higher up. . . . If students at other universities are thinking about doing it, they're going to have to fight to talk to the higher-level management, who might not give them as much time as lower-level administrators.[17]

Going straight to the top led to the creation of the Fair Trade Advisory Board—another opportunity to build bridges. According to one participant, culinary manager Julia Engstrom, "The really friendly nature of our Friday afternoon meetings has made it a great experience. And the inclusion of the people who could help at the bookstore, and the dining director, and the purchasing people, the real estate people. . . . They are willing, energetic, and sincere."

Of course, since 2008, OEOJ students have been building bridges under a huge umbrella. Sustainability coordinator Souder believes that—together—the administration's support, student activism, and connections that link Fair Trade with other sustainability initiatives have led to big changes at UCSD:

> The administration is very supportive of sustainability initiatives and changes, and, in fact, drives a great many of them. They're also very receptive to student programs and interests and passion and initiatives that the students are driving from the grass roots effort as well. So you have a top-down, bottom-up approach that really works well.[18]

Be Persistent

University administrators hear all sorts of requests from students. A successful student-led campaign needs to demonstrate longevity and feasibility.

> Administration is always cognizant that [something a group of students want] has to be applicable in your day-to-day reality. So, yes, it's something they support, and they appreciate the fact that the students are working. . . . If students were out there picketing, saying, "We want Fair Trade coffee and tea and sugar everywhere right now; we don't care what the impact is; we don't care what the cost is; we don't care about the time; we don't care about the availability of Fair Trade products; we just want it right here"—if they were being totally unreasonable, the dynamic would be completely different. But they're working through this carefully and listening—and at the same time, they're not pushovers either. When they see something that doesn't make sense, they question it. It's gotten good support from both sides because of the fact that it's such a cooperative, collaborative effort. Everybody is listening, really truly listening to

one another. In any bureaucracy there are frustrations. And I don't know whether it's just that this particular group of students has negotiated those slow points very effectively and made them not so slow, or whether it's just the excellent cooperation they have from the administration because of their approach and because they've been inviting people to the table to talk and working through the issues, but this I think has moved fairly quickly through the university. They've gotten most of the stakeholders at the table. One way it has affected and changed the way we operate is that every time we work with student groups, or every time we work with administration, we get new tools, and we learn, and we change our approach subtly depending upon the personalities and types of people and disciplines that we're working with, so we've gained tools from this. I think that the approach has changed.[19]

AN ONGOING CHALLENGE

On the UCSD campus there are over a hundred coffee carts, cafés, and other food service establishments that HDH does not run. These privately owned and operated vendors present the biggest obstacle on the path toward becoming a 100 percent Fair Trade University because many of them have long-term contracts that do not stipulate anything about Fair Trade. According to HDH director Cunningham,

> Many of these vendors have long-term contracts, and to make huge changes can be very controversial. . . . To make real change is really hard in thirty-three weeks (an academic year). . . . We have one hour a week to meet, and we're supposed to figure out how to talk to a hundred vendors. HDH is going to carry the flag. Regardless of what happens with these vendors, we're going to try to be the front end of this, because at some point, every movement has to have a champion. . . . With new contracts, we say, "This is the deal." For other vendors, we have kind of a carrot and a stick. I feel very fortunate; the students have been very patient, and they've set very realistic goals to accomplish within the time we have.

For now, the Fair Trade Advisory Board has adopted a policy of encouraging vendors to switch to Fair Trade products and to positively recognizing those who do. There is certainly a role for consumer education here;

student demand is a huge incentive for vendors, who need assurance that Fair Trade will pay off for them. There are already a number of conditions that can make the campus a less-desirable location, such as reduced parking and a tedious proposal process. Without well-articulated demand, it is possible that potential vendors will view Fair Trade as yet another cumbersome requirement for doing business at UCSD.

On this topic, sustainability coordinator Souder comments,

> The students have been very open to saying, "Oh, okay, that's a reasonable constraint that the vendors have. So let's think about how we can reword this to make is something that the vendors can actually do." And they're very cognizant that we all want to push very, very hard, but we don't want to push so hard that we're making it impossible, and that we'll be driving our potential partners away.

LOOKING BACK, SEEING FORWARD

For farmers, Fair Trade is about more than getting a fair price. It is about democratic organizing, exerting more power in global markets, and having real choices in life. We do not often think about empowerment in terms of activism in wealthy countries, but it is not trivial for a college student to see something that she or he wants to change, and to work it through and make that change happen. This is an empowering experience, although one that may include some discouragements along the way. One Earth One Justice founder Linneman told me,

> There was more than one year where we were pretty confident that no one was going to pick up the reins next year. I never thought that [OEOJ at UCSD] would have a life after Viraf—that kid just had energy for days. He's so good at motivating and inspiring people. When he told me, "I attached on to a group of kids, and they're motivated, and they've got good ideas, and they're really moving forward with it," I thought, "Great, maybe we'll get one more good year out of it; maybe we'll get two more good years out of it." Then a couple years later, Viraf forwarded me an e-mail, and said, "I don't even know any of these kids, but these are now the people in charge." I was already comfortable with the fact that the campus chapter was being run by a bunch of people *I* didn't know. But

> when it was then being run by a bunch of people that *he* didn't know, that's when it dawned on me. A few years ago, if you had asked me about One Earth One Justice, I would have told you that it flat-out failed. But it was still gathering speed; I just wasn't aware.

Pioneer OEOJ member Seymour was, likewise, pleasantly surprised,

> UCSD now is doing what we started. The principle of Fair Trade is about purchasers' power to vote with our money. It's really funny, because I think back to that time when six or seven of us were first meeting, figuring out how we were going to do this. If you had told us that Housing and Dining would be putting UCSD's purchasing power behind Fair Trade, I'd have thought you were crazy!

Claire Tindula, a 2009 graduate, reflected,

> I was just talking to Viraf, and he was saying how, when they started the movement, they didn't really envision it going this far, even with their optimism and with helping people to get involved, but the goal that we've always talked about is getting Fair Trade from being this specialty niche market to being the standard, like "Why *wouldn't* you have Fair Trade?" So, with the Fair Trade University, which has come about because of the work of OEOJ, but also has been supported by an increasingly knowledgeable student body, we've gotten to the point where Fair Trade is not just a "flavor."

One Earth One Justice member Chris Westling saw a learning curve:

> [Two of us] got into A.S., and that didn't really give us power, but it gave us the power of knowledge, and so we started figuring out the system. OEOJ has been around for four or five years, and the first four years we could have done—knowing what we know now—in a week. But it would probably be different if someone else had told us; we've learned it on our own.

Of course, it takes even the cleverest and most experienced activists much more than a week to build an organization and a movement. What Westling learned is that to change something about an institution, one must first understand how that institution works.

CONCLUSION

There are several reasons why Fair Trade campus campaigns can be very effective. First of all, Fair Trade, although political in its underpinnings, is a "feel-good" cause. Second, the movement has an established history. It is not a fad but a change in lifestyle and behavior. Third, Fair Trade raises awareness of a problem and offers a solution; it is relatively easy for consumers—ever poor college students—to understand why spending five cents more for a cup of coffee or tea makes a difference in the lives of the people who grow these crops. And finally, there are many targeted resources available via Fair Trade organizations, such as Global Exchange and United Students for Fair Trade, and most of the country-specific labelers.

In the UCSD case a convergence of students and university employees united under the umbrella of a serious sustainability initiative, but I do not believe that such an overarching vision is a necessary precondition for a Fair Trade University. Rather, I think this case suggests a strong reason for Fair Trade activists to *frame their message as an important part of a larger picture of global social, environmental, and economic sustainability*. And, as with any sustainability effort, the goals and strategies are constantly evolving through evaluation and refinement. Becoming a Fair Trade University is a beginning, not an end point.

From Niches to Norms

At UCSD, Fair Trade is part of an institutionalized effort to change norms about consumption, waste, social justice, and respect for the planet. Some of the university employees I interviewed still see Fair Trade in terms of offering people choices, that is, as a flavor—a product line that caters to a niche market of socially conscious customers. Eventually, consumers on campus will not have to make decisions about human rights or environmental protection every time they buy a beverage; the choice will have been made for them. But even when that point is reached, activists see education as a very important component of the Fair Trade University. Just as recycling has a more positive impact when it accompanies conscious consumption, choosing Fair Trade—or having it chosen for you—is most powerful when it is backed by or leads to an understanding of the larger issues around global trade inequalities.

Asked about the connection between buying Fair Trade products and supporting other, related, transformative agendas, such as antisweatshop campaigns, trade policies that promote human rights and environmental protection, or reforms to the US Farm Bill, OEOJ members responded:

> There are people out there who buy Fair Trade because they're conscious, but I'm not sure to what extent *all* the people who buy Fair Trade products are conscious. There's a whole, embraced system of consumerism. So we've got people buying based on pricing; we've got people buying based on whatever—maybe organic. So I like to compare it to the organic movement. When you talk to some people, they are very aware of pesticides and their environmental effects. But very rarely are people that passionate about the social issues. They know Fair Trade is good and they should be doing it, but they're not activists, they're not politically charged. And that might not be such a bad thing. It's just that when your goal is a human social rights and justice issue, that's what Fair Trade is. But by creating a consumerist system where selected products are Fair Trade, it places less focus on the fact that all trade should be fair.[20]

> Most of the people I've spoken to, even just in the campus community, are aware of the issues. What's really encouraging is that, when I mention Fair Trade, and people who don't know what it's about ask for an explanation, and a lot of times once I explain a little more about the certification they are very excited because it touches on all different aspects of labor issues and trade justice and environmental sustainability. If people don't already know about Fair Trade specifically, they are aware of the issues surrounding it and what it impacts, so when they hear about Fair Trade, it's within a context of a larger awareness. . . . Some people will buy Fair Trade products as part of their shopping routines, but for those who are more politically active, Fair Trade can become a force in alerting them to bigger issues and making them more active on them. It can be a mobilizing force.[21]

> I can't imagine that people would stop with paying an extra couple of cents more for Fair Trade. They feel better about themselves and what they've done, so they want to do more. I know that's what happens with me.[22]

> Within OEOJ, we joke that we're the most aggressive pushers for Fair Trade on this campus, and at the same time we're the most anti–Fair Trade people because we recognize that Fair Trade is just a step toward a market solution helping to alleviate certain inequalities in poor farming communities. But it is ultimately a market-based solution. It doesn't speak to the larger, more structural issues of inequality in the global trade system; it just corrects a few relationships in the chain. We have to figure out a new system of global trade. So we recognize that [the Fair Trade movement] might be delaying a wake-up call. . . . It doesn't ask how or why we claim that capitalism is such a great system and that the free market is going to save everyone. If that's the case, why do we need these movements in which people are working for free to act as a corrective mechanism? But we still recognize that, for the farmer who's producing the coffee, it makes a difference. And seeing a logo again and again, someone's going to at some point say, "What does that mean?"[23]

These comments are realistic yet hopeful. Some consumers will put their dollars behind Fair Trade because it makes them feel good, or just because it's there, and this helps farmers in the global South just as much as when someone makes a more conscious, politicized choice to buy a Fair Trade product. But the students also strongly suggest that, over time, exposure to Fair Trade gets people thinking; it can light a spark of interest in the bigger picture of how one's consumer behavior matters and also an interest in social, environmental, and economic sustainability at the policy level.

The visibility of Fair Trade provides an ethical reference, raising the standards of social responsibility by which businesses or institutions are judged. The more consumers know about international trade issues, the more important this normative ethical bar becomes (Gendron, Bisaillon, and Rance 2009). For a Fair Trade University, this means constant a reevaluation of its practices—not only regarding products sourced from distant and usually poorer places but in terms of how the university, through its purchasing power, can encourage sustainability writ large via its relations with all sorts of suppliers, contractors, and workers.

Chapter 7

Growing Fair Trade

To be sure, Fair Trade is growing. Fairtrade International's 2009–10 annual report documents that, by the end of 2009, there were 827 certified producer groups worldwide and more than 2800 companies licensed to use the Fair Trade label on products. Between 2008 and 2009,

1. The estimated retail value of Fair Trade products rose 15 percent to almost 3 billion (about $4.1 billion).
2. Sales grew by 25 percent or more in eight countries.
3. Despite a continuing recession, no markets fell back.
4. Sales of Fair Trade sugar, fruit juices, and herbs and spices more than doubled.
5. Almost twelve million liters of Fair Trade wine were consumed—an increase of 33 percent. (FLO 2010a)

Fair Trade USA releases reports that focus not only on sales volume but also on increased income to farmers. Table 7.1 documents US imports of Fair Trade–certified products from 1998–2009. In 2010, US sales of Fair Trade products accounted for over $14 million in social premiums (Fair Trade USA 2011b, 14). The proliferation of Fair Trade Towns, which did not even

Table 7.1
Imports of Fair Trade Certified Products into the United States, in Pounds, 1998–2009

Year	Coffee	Tea[1]	Cocoa	Rice	Sugar
1998	76,059				
1999	2,052,242				
2000	4,249,534				
2001	6,669,308	65,261			
2002	9,747,571	86,706	14,050		
2003	19,239,017	95,669	178,888		
2004	32,974,400	180,310	727,576		
2005	44,585,323	517,500	1,036,696	73,824	271,680
2006	64,774,431	629,985	1,814,391	390,848	3,581,563
2007	66,339,389	1,134,993	1,951,400	436,456	8,657,427
2008	87,772,966	1,372,261	3,847,759	317,652	8,696,172
2009	109,795,363	1,372,157	2,629,411	971,453	10,963,627
Total	448,275,603	5,454,842	12,200,171	2,190,233	32,170,469

1 2001 tea figure includes tea certified in the second half of 2000.
2 Includes avocados, bananas, citrus, grapes, mangos, and pineapples.
3 Unit measure for flowers is stems.
4 Unit measure for wine is liters.
Source: TransFair USA (2010)

exist in the United States before 2006, also evidences the success of Fair Trade activists' efforts.

Moving forward, how can the Fair Trade movement make the most of its accomplishments? Fair Trade is growing by certifying more commodities and reaching into new markets, but internal tensions accompany this

Year	Produce[2]	Spices	Flowers[3]	Honey	Wine[4]
1998					
1999					
2000					
2001					
2002					
2003					
2004	8,814,171				
2005	7,384,202				
2006	6,176,907	197,145			
2007	8,030,482	149,460	650,832		
2008	25,492,767	44,165	9,835,028	266,385	193,518
2009	50,272,722	149,344	9,539,859	242,671	1,450,717
Total	106,171,251	540,114	20,025,719	509,056	1,644,235

expansion. This chapter summarizes efforts to expand Fair Trade's reach and the challenges the movement is facing. In its conclusion, I—with the help of others—argue that Fair Trade is indeed promoting norm change on several different levels, and that different visions of where the movement is headed need not be at odds.

NEW PRODUCTS, NEW MARKETS

Maya Spaull's job at Fair Trade USA is to introduce new Fair Trade Certified product categories in the US market, and to poise these products for success. In recent years she has helped bring Fair Trade Certified teas, herbs and spices, honey, nuts and oils, grains, and wines to American consumers.

> I work to diversify Fair Trade offerings because the more products and markets we have, the more farmers and farm works in the global South we can reach. In January 2009 TransFair USA opened up to everything FLO [now Fairtrade International] certified. That was a change for us. Initially our strategy was to launch products strategically and in categories in which we could have the deepest penetration for producers and in the market. But then we realized that it was time to just harmonize with our international partners and allow the opportunity for any producer who has obtained Fair Trade certification to sell into our market. And that's been really nice. We've seen new products come in, from olive oil to quinoa, just within the last year.[1]

To successfully introduce a new category of Fair Trade products in the United States, Spaull must first consider feasibility. She talks to distributors and retailers, and looks at where the producers are.

> What is the need for Fair Trade? How can Fair Trade certification make a positive impact? How much demand is there for the product? Is there adequate supply of the Fair Trade Certified product so that if the U.S. market demand was there, that we would be able to certify it? What is the minimum price and Fair Trade social premium; is that bearable? We look at the characteristics of a product as it exists in the industry now. So for wine, what are people buying? Why are they buying it? Who's the target consumer particularly for sustainable wines, for organic wines? We also look at the ease of certification. Does the product have a very complicated, multistep supply chain, or is it a finished product like wine that comes here already Fair Trade labeled in the bottle—a little easier to certify?

Answering such questions helps Spaull and her team decide whether or not their efforts to introduce and promote a new Fair Trade product will be "worth it" in terms of making a positive difference for producers. If they decide to move forward, they identify potential industry partners and pilot products, and build slowly from there. Says Spaull,

> Because there are a lot of resources that go into developing a new category, we do a pretty good job of ensuring that the opportunity for the product is substantial and diversified. While we'd love to work with every producer that needs our support, we have to work with the products that are going to affect the most producers and have the most ultimate impact.

Fair Trade Textiles

More recently, Fair Trade has extended its impact by testing certification for apparel and linens. In 2010, Fair Trade USA published standards for a two-year pilot project that now includes twenty clothing brands and four factories located in Costa Rica, India, and Liberia. A Fair Trade label on a shirt, for example, means that the shirt was made in a factory that is committed to upholding internationally recognized labor standards and to the well-being of its workers. Such factories are held accountable through inspections and grievance processes. The shirt's brand, in turn, has agreed to pay a Fair Trade premium to workers, to engage in a longer term business relationship with the factory and is paying prices that help the factory consistently uphold Fair Trade standards.

The core factory standards apply the worker empowerment elements of Fair Trade to apparel and linen manufacturing. They establish minimum workplace conditions based on conventions of the International Labor Organization and incorporate best practices provisions set out by global, multi-stakeholder initiatives such as Social Accountability International (SAI) and the Worker Rights Consortium. There are specific standards pertaining to transparency, worker participation (including a functional grievance process), Fair Trade management systems, the Fair Trade premium, forced labor, collective bargaining, health and safety, child labor, wages and benefits, working hours, the employer-employee relationship, non-discrimination, disciplinary practices, environmental management, and women's rights.[2] At the time of this writing, seven factories have applied for certification and four have qualified.[3]

To become licensed to sell Fair Trade clothing or linens, brands must use Fair Trade cotton (for cotton products), produce garments in factories approved by Fair Trade USA, purchase products according to Fair Trade terms, and pay a social premium directly to a worker-controlled fund (Fair Trade USA 2010). For the pilot project, Fair Trade USA has worked with "small mission-based brands with lots of credibility and experience promoting sustainability and better working conditions."[4] One of the first brands licensed to sell Fair Trade apparel was Maggie's Functional Organics. The company's apparel chain includes organic cotton growers, knitters and dyers, and cutters and sewers—all of whose wages and working conditions are governed by Fair Trade standards (Maggie's Functional Organics 2011). Another pioneering brand of environmentally conscious clothing is prAna. Since March 2011 their popular "SOUL T" carries the Fair Trade label and is creating jobs for women in post-conflict Liberia (prAna 2011).

Factory auditing initiatives have been in place for some time now, with disappointing results.[5] Fair Trade Certification offers a broader and more sustainable approach to ethical clothing. It is a "carrot" for brands that want to appeal to socially conscious consumers and for factories that want long-term contracts with buyers who pay enough to actually support fair wages and good working conditions. The Fair Trade social premium is an additional benefit directly to workers that increases their take-home pay or supports investment in their communities. To date the apparel and linens project is still quite new and encompasses only a handful of manufacturers and brands, just as was true of Fair Trade certified coffee and other food products in the early years. Larger brands are taking a "wait and see" approach to certified clothing; their participation will hinge on an increase in the number of approved factories and confidence in customer demand.

An Alternate Certification

In 2006, the Institute for Marketecology (IMO, corresponding to the organization's German name) introduced the Fair for Life label. Its criteria are based on the International Labour Organization (ILO) conventions, FLO Fair Trade standards, the SA8000 social accountability standard,[6] and International Federation of Organic Agriculture Movements (IFOAM) principles. Fair for Life is now a prominent label in the US, appearing on textiles and body-care products as well as coffee and chocolate bars. Some producers prefer this certification because it is somewhat more flexible than Fair

Trade in that it can be applied worldwide, to manufacturers as well as growers, and encourages direct negotiations between sellers and buyers to determine prices and social premiums. It allows for certification of multi-ingredient products such as sauces, infused oils, soaps, and cosmetics by using content rules modeled after those for organic products. Like Fair Trade USA, IMO considers the whole value chain when certifying textile products.

7.1 Fair for Life Label

Fair Trade Ingredients

Cate Baril, Fair Trade USA's director of business development for grocery and ingredients, works to expand the market for Fair Trade ingredients as *components* of packaged goods such as energy bars, cappuccino mix, vanilla soymilk, ready-to-drink iced tea, and—of late—soaps, shampoos, and lotions. This strategy carries huge potential for growing demand for Fair Trade Certified products, because some of the costlier ones, such as organic sugar or extra virgin olive oil from Palestine, can be used as ingredients without prohibitively raising the cost of the finished candy bar or bar of soap.

7.2 Fair Trade Ingredients Label

Part of Baril's job is to help companies explore how they are purchasing the components that go into their products and how else they could be purchasing. She links Fair Trade ingredients to the larger sustainable foods movement.

> People don't see farmers in the middle of the grocery store. When you see an apple, you see a farmer, when you see a box of cereal, probably not. Fair Trade may help people make the link. . . . We have so many close ties to so many other places, there are so many stories. When the brands can understand this, there's a lot they can do. It's the same with sourcing local foods.

Besides working with "the brands"—companies that manufacture sustainable food and body-care products—Baril and her team go directly to retailers, educating CEOs about Fair Trade and asking them to follow up with their buyers, who in turn can influence suppliers. Whole Foods

Box 7.1

Ingredients of Sustainability

A CONVERSATION WITH JEREMIAH MCELWEE, WHOLE BODY COORDINATOR, GLOBAL PURCHASING TEAM, WHOLE FOODS MARKET

Jeremiah McElwee coordinates Whole Foods Market's twelve regional teams that oversee the purchasing of body-care, nutritional supplements, and lifestyle products. He "helps people connect the dots" by, for example, creating new efficiencies, introducing new products, and sharing data. McElwee describes Whole Foods Market as a "mission driven" company with high standards regarding the socioeconomic and the environmental impact of the products it sells.

How does one know if a body-care product or nutritional supplement product is sustainably sourced?

There are obviously a lot more layers than with a single-ingredient item such as coffee. Whole Foods Market team members ask a lot of questions. Sometimes we ask things a supplier has never been asked before. Such as how are the ingredients processed? What are your starting materials? What is the country of origin of the ingredients you have selected?

How sustainable is "sustainable enough" for Whole Foods Market?

There's not one simple answer to this. We have longstanding, trusted relationships with many of our suppliers. We also have seasoned buyers who know what sorts of questions to ask. With newer suppliers it comes down to a system of "organic vetting" by

now has at least twelve hundred Fair Trade products in their system, and Wal-Mart is asking for Fair Trade. There is a tremendous challenge in supplying such a large company, "but the amount of people whose lives this can change, instantly, is amazing. When a company we work with has the desire, and the scale, to sell to Wal-Mart, we try to help them."

Wal-Mart is a touchy subject within the Fair Trade movement. Baril acknowledges this, but points out that Fair Trade is a certification, not a brand.

> You can walk down the tea aisle in Whole Foods and practically all the teas are Fair Trade, but the brands are really different. Fair Trade is just one attribute of what the brands offer. And I think there's room for all of

teams of buyers and our quality standards team. Our industry is tightly knit, so it acts as a network that is self-policing by nature. IMO Social Responsibility and Fair for Life, Rainforest Alliance, and Fair Trade USA help with defining sources and certifying fairly traded ingredients. Then Whole Foods Market can go to a supplier and ask them to use these certified ingredients, hence putting Fair Trade on suppliers' "radar."

Are there other partners in this effort?

We work in collaboration with countless mission-driven suppliers in the natural products community. All share an understanding of issues, but each has a slightly different passion or area of expertise. Another kind of networking is to connect the "old guard" and the "new guard" within the industry. The "old guard" or early founders of companies have experience that can be relevant to new companies starting out. The "new guard" has ideas, and possibly a more current understanding of sustainability issues, as well as links to new forms of communication like social networking. We also network outside of retail, farming, and manufacturing, with, for example, Science Nature Art People (www.snapgathering.com) and microcredit NGOs.

What's the bottom line for you?

If we want to maintain a capitalistic system, where we all can find products we love and can purchase them, we need to be transparent about supply chains from cradle to cradle: sustainable farming practices, responsible packaging, and awareness for consumers on all aspects of externalized costs associated with the goods they purchase.

Telephone interview by A. Linton, August 25, 2009, and e-mail correspondence on January 28, 2010.

them. I know there definitely are groups who would like Fair Trade to be completely for truly just the most marginalized of farmers, but I think we need to figure out how to affect as many farmers as possible.

Furthermore, brands that answer demand from quality- and sustainability-conscious consumers are not impacted by what lower-end brands do. For example, the market share of a high-end brand of Fair Trade coffee will not be affected by the introduction of a less distinctive Fair Trade coffee at Sam's Club (a division of Wal-Mart), which can be sold for less than half the high-end price because it was roasted, ground, and packaged in Brazil and brought to market by the container. The farmer cooperatives that grew the coffee still received the Fair Trade price and social premium.

I asked my UCSD interviewees for their thoughts on Fair Trade products in big-box stores. Here One Earth One Justice member Yuki Murakami weighs in:

> This is something I've struggled with personally. On one face of it, I think to myself, if Wal-Mart is bad to their own workers and they try to cut down hours so that they don't have to pay full-time wages or provide full-time benefits, and they don't really look at the communities they are destroying, then why should we sell them Fair Trade? If that's the idea, where do you start with Wal-Mart? Do we tear them down, or do we make one concession so that a lot of people who want this can live better and Wal-Mart might actually have kind of first step in the right direction? So, that's the very, I guess, optimistic view of it, like, one, let's give 'em a chance or two. It's making a small difference, so it's worthwhile.

GROWING PAINS

Does it matter whether the penultimate link in a Fair Trade value chain is a World Shop, a Starbucks, a Tesco, or a Wal-Mart? In terms of benefits to producers, sometimes the answer is "yes" and sometimes "no." Recall from chapter 4 that farmers who supplied Fair Trade fresh fruit to large supermarket chains found themselves at a disadvantage, whereas other farmers who supplied Fair Trade cocoa to the same chains profited (Barrientos and Smith 2007). Another study showed that, regardless of a company's reasons for stocking Fair Trade products, the advantages to producers were similar (Bezençon and Blili 2009).

Sociologist Laura Raynolds (2008) distinguishes between three types of businesses that source Fair Trade products: "mission-driven" companies that actively support and promote all aspects of Fair Trade, "quality-driven" buyers for whom Fair Trade is one strategy to ensure a reliable supply of premium coffee, and "market-driven" businesses which may (or may not) have progressive social responsibility policies but are advancing mainstream business practices that foster competition between suppliers within buyer-dominated value chains (i.e., the buyers call the shots). While Fair Trade relationships with market-driven buyers have vastly increased revenues to producers, these relationships are often less stable than other Fair Trade partnerships. This is largely because large corporate buyers are reluctant to provide prefinancing. Producer groups know that buyers may refuse to buy from them if they ask for the prefinancing they are entitled to under Fairtrade International's guidelines. In addition, producers' relationships with market-driven buyers are not as long-term as those with mission-driven and quality-driven buyers. Still, because of their size, market-driven companies may still be putting more money into Fair Trade value chains than mission-driven or quality-driven businesses are able to. For this reason I argue that, *to the extent that all actors abide by the rules of Fair Trade,* both niche and mainstream buyers are promoting more ethical trade relationships when they source Fair Trade products.

In her study of transnational activists, political scientist Shareen Hertel observed that "members of the tightly knit networks of people involved in international campaigns can ostracize as effectively as they can include one another. Longstanding political differences, personality clashes, betrayals, and dashed expectations are just as much a part of transnational advocacy as international brotherhood and sisterhood" (2006, 28). There is certainly evidence of such tensions within the Fair Trade movement due to disagreements over strategy, powerful egos, unmet expectations, or simple misunderstandings. But there is also room for accommodation, if not consensus. Discussing controversy over mainstreaming standardized Fair Trade criteria versus promoting more stringent certifications, such as Bio-Équitable, which combines organic and Fair Trade standards, development scholar John Wilkinson argues that mainstreaming does not represent a moral threat to Fair Trade as a movement but should rather be understood as one of its strategic components. "In important ways, the opposed wings of the movement are reinforced by the activities of others" (2007, 231). He points out that

> Fair Trade is currently "on a roll" with its legitimacy high both in the market place and the State. Although the latter element is not found in the U.S., public sector involvement should not be seen as a European peculiarity since it also extends to the UN system. . . . Many in the movement and much academic analysis have focused on the dangers of mainstreaming and it is clearly a risk for the independent image of the movement. Some, however, would see mainstreaming as the ultimate goal of Fair trade with its standards being adopted as the bottom line for all trade. (236)

In a similar vein, sustainability scholar Corinne Gendron and her colleagues acknowledge that Fair Trade value chains incorporate two distinct distribution channels: an alternative commercialization network of mission-driven businesses and a pathway for Fair Trade products to be identified and sold in the mainstream marketplace. They argue that "if the Fair Trade movement really wishes to support Southern producers, it must focus on volume, without which results remain marginal and insignificant" (Gendron, Bisaillon, and Rance 2009, 69).

Gendron suggests that mainstreaming per se should not be Fair Traders' primary concern. The real issue is how to mainstream Fair Trade in a way that is acceptable to those who see the movement more in political than in economic terms. Those who prioritize the political see Fair Trade as a necessary "Band-Aid" that helps producers in the global South reap the rewards of their labor while activists forge ahead, pressuring governments and intergovernmental bodies (ultimately the World Trade Organization) to institutionalize trade rules that mandate economic justice and social and environmental sustainability. For example, sociologist Marie-Christine Renard argues that more coffee being sold under Fair Trade terms does not guarantee that all actors involved will adhere to the ideological principles of the movement. It is "vital not to lose sight of the social interactions on which Fair Trade was built and which legitimize it, and of the importance of mobilizing them" (2003, 96). Further, Renard (2005) criticizes the Fair Trade system for not offering farmers incentives for crop diversification and for not adjusting the "fair" price in correspondence to national or regional economies.

Another strong and vocal proponent of Fair Trade as a global political movement is farmer/priest/scholar Francisco VanderHoff Boersma of Mexico's UCIRI coffee cooperative. He thought-provokingly posits that Fair Trade is not primarily about poverty reduction but rather about embedding the economy in society by creating *a different sort of market* on foundations of quality and efficiency, ecological sustainability (including crop diversification and food security for farmers), and social sustainability—integrating the entire cost of production into products and reducing social distance between producers and consumers. VanderHoff Boersma argues that "the desire to 'help the poor' can easily divert Fair Trade organizations from following the rules of the different type of market established by small producers and can thereby undermine the long-term prospects for extending this market" (2009, 58). Thus, activists need to focus on changing the system, not on fixing an effect of the system (poverty). But, provocative prose notwithstanding, VanderHoff Boersma does not provide a single example of how using Fair Trade to get more money to farmers is undermining the long-term prospects for social change that he and his fellow farmers envision. And, as long as the Fair Trade terms of contract and payment are upheld, where the money is coming from (e.g., a Wal-Mart or a World Shop) is not as important as how producers are using it to achieve sustainable livelihoods and to contribute to their communities—both of which can and do involve transformative agendas initiated at the local level.

Those who focus on spreading Fair Trade's economic benefits to as many people as possible certainly do not eschew the goal of a world where all trade is fair on all dimensions, but their approach is more pragmatic and immediate: make a difference in people's lives now. From this perspective, the thing to be cautious about as Fair Trade grows is *integrity*. Rick Peyser, Green Mountain Coffee Roasters director of social advocacy and Coffee Community Outreach and Fairtrade International board member, expresses it thus:

> Maintaining the integrity of the seal while going to scale is the challenge now. Don't just look at the product and the market demand for the product. What is Fair Trade doing for the farmer? The movement needs to be sensitive. Who are we trying to benefit? Are we here to support a business model in the North or producers in the South? The business model has to get behind supporting the farmer.

Peyser does not talk politics, but he does foresee social transformation as more and more businesses act in their own enlightened self-interest. "Some day [Fair Trade Certification] will no longer be necessary. There will be enough knowledge and understanding about the challenges that companies will set prices around the long-term health of their supply chain."

Representing Fair Trade USA, Spaull elaborates upon a similar vision:

> I think that Fair Trade is an innovative and a different way to support business as usual. There's such a power in activating consumers and industry to do what they do already in a different way. So I would like to say that, eventually, if we do our job well, and there's enough consumer awareness and industry transition that "Fair Trade" practices will be busi-

> ness as usual. Workers will have safe and fair conditions, and they will be paid well. Farmers will be receiving a price that is livable—that covers their cost. Overall, there will be a more level economic playing field globally, and we are a tool to help people get there. There's a lot of power in third-party certification. Industries are increasingly concerned about their corporate social responsibility programs and their mission, and how to incorporate that into their bottom line. So I think there's always going to be a place for the foreseeable future for certification. We are very excited about the new directions that Fair Trade certification can go in, from seafood and fisheries to manufactured goods, starting with garments, and perhaps eventually toys, electronics, and computers. Fair Trade USA is also researching how to potentially apply the Fair Trade model to domestic US production. We are interested in helping address the labor issues that are problematic here on farms, and supporting the changing needs of American small family farmers.

By now it should be clear that we are not really talking about contrasting visions but rather about divergent strategies. Still, activists (and, to some extent, mission-driven businesses and conscious consumers) do what they do because they are very dedicated to a specific ideal. This is why Gendron, Bisaillon, and Rance seriously address a need to ideologically reconcile the "political versus economic" debate. They do so by showing how the economic side of the Fair Trade movement is helping the political side at the same time that it is increasing revenues to farmers and their communities. Responsible consumption is a very efficient means to state one's beliefs, considering that we in the global North consume every day but only vote once in a while. It also "influences practices beyond the national jurisdictions to which a citizen is subjected, thus presiding over a real globalization of social mobilization" (2009, 72). This does not imply that voting with one's wallet is an alternative to traditional political actions such as voting in elections, communicating with one's elected representatives, or protesting, but rather that it is one of several strategies that can work together to effect social change. Conscious consumption is "a means for citizens to construct political action within the market where *unsatisfied demands on the political system can be expressed in the era of globalization*" (72, italics added).

Furthermore, Gendron and her colleagues remind us that "the existence of Fair Trade itself serves as an ethical reference, fixing the standards of social responsibility by which large businesses and their operations are now judged" (73). Ten years ago it was rare to find sections about sustainability or social responsibility on large, mainstream companies' websites or in their reports to shareholders; now these are ubiquitous. The statements often lack credibility or verifiability, but the point remains that Fair Trade, anti-sweatshop, and other new social movements grounded in political consumption have made it very difficult for corporations to simply ignore or hide their impacts on people and the environment. Some offer sweeping and largely unsubstantiated claims about their commitment to sustainability, but this is not difficult for conscious consumers to recognize. For example, Kraft Foods' "responsibility" web page, which is linked to their "corporate" site aimed at investors, tells readers that the multinational food giant is serving needy communities ("Over the past 25 years, we've donated more than $770 million in cash and food") and follows a code of conduct based on obeying the law and making decisions that are ethical for "our business, our shareholders, our employees and our consumers" (Kraft Foods 2010). Consumers cannot wrap their heads around what "$770 million in cash and food" over twenty-five years has amounted to, nor do they have any way of knowing who these donations have benefitted. Likewise, they cannot read Kraft's code of conduct for themselves, or learn how compliance is documented or noncompliance is punished. Kraft recently purchased Cadbury Chocolate, and has agreed to honor Cadbury's Fair Trade sourcing commitments but not to increase the amount of Fair Trade cocoa in the Cadbury line (Carrell 2010).

The "sustainability" page of Wal-Mart's website begins:

> At Wal-Mart, we're now using the sun and wind to help power our stores. Our drivers' routes are shorter, which means less diesel fuel and greenhouse gases. And the way we pack our trucks is far more efficient. Why do we do it? For one thing, it makes a difference for the environment. But we also learned these efforts save us money—a savings we can pass along to you. Which means we're not just being earth-friendly, we're also being wallet-friendly. How great is that?

Wal-Mart's page contains links to information about "earth-friendly" products that the company sells as well as tips about reducing waste, conserving water, eating organic fruit, and cutting down on pollution created at home, for example, by properly disposing of unwanted chemicals and switching to nontoxic cleaning and garden products. To the extent that such information encourages Wal-Mart shoppers to, say, switch to compact fluorescent light bulbs or shop with reusable bags, it is promoting sustainability. But the site does not provide the specific information that one would need to assess the magnitude or effects of the sustainability measures the company touts. And noticeably absent is any discussion of sustainability in terms of the livelihoods of the workers who grow, sew, or manufacture the products the company sells—or in terms of its own employees' well-being. These shortcomings can (and should) be read as cleanwash, greenwash, or (in the Kraft example), bluewash. Conscious consumers readily notice the opacity of statements such as Kraft's and the inconsistency of Wal-Mart's "sustainability" message; they may well decide to avoid purchasing Kraft products, to not shop at Wal-Mart, or to engage in organized actions to persuade these firms to improve their practices. But would anyone argue that Kraft should cease to address ethical sourcing, or that Wal-Mart should stop selling compact fluorescents or eco-friendly cleaning products?

One cannot deny that, even if new social economic movements like Fair Trade may produce less spectacular transformations that an improbable revolution of international commercial rules and worldwide economic governance, they are no less capable of provoking transformations. As geographer Julie Guthman somewhat grudgingly acknowledges, Fair Trade labels "are the only existing food labels that are clearly redistributive in their intent, and perhaps in their practice." She goes on to note that "protective labels" such as Fair Trade necessarily create barriers to entry; only some producers can qualify (2008, 205). Thus, they are no panacea to the problem of exploitative trade relations. However,

> if labels are one of the few tools available to mitigate the injustices and destruction of neoliberalization, it may be worth saving the bathwater over the baby in this case. Because although the labels themselves may be limited in their positive effect and even produce perverse outcomes[7], their saving grace may be in their unintended consequences. (206)

One "unintended consequence" of ethical labeling is that

> companies must now work with a definition of their social performance fixed not only by their public relations specialists but also by myriad

> NGOs comprising both Southern actors and Northern activists. The new social economic movements also participate in an ethical restructuring of the market, such as Fair Trade, through which the treatment of Southern producers can be used as a commercial differentiation factor. Trade logic and social responsibility can thus converge while supporting strategies for more political institutionalization and reinforcing the process of transforming the rules of international trade. (Gendron, Bisaillon, and Rance 2009, 75)

MOVING AHEAD

As the previous section has illustrated, Fair Trade activists have many grounds for agreement and partnership based on improving livelihoods, protecting the Earth, *and* changing the norms of global trade. Looking forward, VanderHoff Boersma emphasizes that *producers must maintain complete control over the chain of production and be the primary beneficiaries of their labor.* Specifically criticizing a decision to allow Nestlé to source Fair Trade products because the company's "practice runs completely counter to Fair Trade values," he implies that dealing with such buyers automatically undermines producers' control and benefits (2009, 58). This need not be the case. I certainly do not wish to dissuade activists from pressuring transnational corporations to better their practices, but there is a powerful role for Fair Trade to play in helping them do this. The best way to responsibly grow Fair Trade, with producers maintaining control and reaping their just rewards, is to let certified producers decide to whom they want to sell, just as consumers decide where (and if) they want to buy Fair Trade products.

It follows that the main imperative for Fair Trade as it expands is *transparency*—especially in terms of producers' access to information about buyers and consumers, something the case studies in the first part of this book indicate is sorely lacking (see also Goodman 2004, Wilkinson 2007). I am much less concerned about some Fair Trade producers' scant understanding of the trading system in which they are involved than I am with their imperfect knowledge of who the buyers are, the terms of their contracts, and their products' final destinations. Without such knowledge, producers cannot make informed decisions regarding firms they do, or do not, want to do business with. Of course it is not possible for small producer groups that sell to Fair Trade importers to know exactly where their produce is going because these importers supply many different mission- and quality-driven buyers, but this is not a problem. Certainly producers should be able to accurately weigh the costs and benefits of dealing with large firms that import directly. *The first priority of the global Fair Trade movement should be to create a window through which producers can see buyers at least as well as buyers can see them.*

This is a time of rapid and deep change in the Fair Trade movement; it has become quite apparent that there is neither a single Fair Trade system nor a unified way to pursue social justice through trade. But diversity can augment strength. On September 15, 2011, Fair Trade USA and Fairtrade International (FLO) issued a joint statement announcing their separation. They officially parted ways on December 31, 2011. Fair Trade USA has three ambitious goals for growing Fair Trade:

* Strengthen farming communities via credit, capacity building, and development.
* Innovate to make Fair Trade more inclusive and scalable, with a target of doubling impact by 2015.
* Ignite consumer demand through education.

In pursuit of these goals, Fair Trade USA will continue to work with producers who have FLO certification, but in addition will offer an alternative certification via a reputable organization, Scientific Certification Systems. This will make it possible for plantation-grown coffee, sugar, and cocoa to bear the Fair Trade label in the United States and for plantation workers to benefit from a system from which they had been excluded. To implement their "Fair Trade for All" strategy, Fair Trade USA will partner with other NGOs including some that are promoting domestic fair trade, notably the United Farm Workers of America.

As has always been the case, Fair Trade USA will not bind producers or businesses into exclusive relationships with them. Producers own their certifications and can sell to anyone they choose. Likewise, businesses are free to work with Fair Trade USA as well as FLO and other certifiers such as IMO and Rainforest Alliance.[8]

Both Fairtrade International and Fair Trade USA maintain a commitment to increasing Fair Trade's impact on producers' lives, and thus both emphasize innovation. FLO is constantly revising standards to better reflect producers' realities and needs, but the international membership does not support certifying plantations that could then compete with small-farmer cooperatives within Fair Trade markets. In addition, Fairtrade International strongly believes in a global Fairtrade label, something that Fair Trade USA has resisted. The two groups are working together to effect a smooth transition for all stakeholders (FLO 2011b). This development opens up new prospects and challenges for producers, businesses, and researchers. Will competition between plantations and cooperatives really be a problem, as feared by some, or will markets expand, creating a significant upsurge in demand for Fair Trade products? In the United States, will businesses predominantly choose to use Fair Trade USA as a "global service provider" for an expanded supply of Fair Trade Certified products and links to Fair Trade towns and universities? Or will they work with several different labelers? Should scholars make a distinction between producer groups that work with Fair Trade USA and those who do not? How can we best measure the impacts of Fair Trade USA's new strategy compared to FLO's constantly evolving strategies? These questions merit further consideration in the coming years.

The future of Fair Trade also involves a holistic vision integrating producers from afar, the family farm up the road, and—more generally—conservation, alternative energy, and waste reduction. As demonstrated in chapters 5 and 6, a growing number of Americans are already making this connection—individually and via their businesses, communities, or institutions. Europeans may be wondering what took us so long. Some progressive examples of integrated Fair Trade can also be found in countries that are home to both producers *and* consumers, such as Mexico and Brazil (see, e.g., Jaffee, Kloppenburg Jr., and Monroy 2004; Wilkinson and Mascarenhas 2007). These initiatives hold great potential for linking Fair Trade certification, local farmers, agricultural workers, and conscious consumers—including those of limited means. But perhaps, because the United States has seen decades of opaque producer-consumer relations in which—with some exceptions spurred by labor union campaigns—people did not care where their food (or clothing) came from, the US case most clearly shows the power of Fair Trade as a conduit via which individuals and institutional actors become aware of international trade issues, such as unfair and unsustainable distribution of wealth, environmental degradation, and social disruption. At least for some, this awareness is translating into purposeful action.

Fair Trade will continue to play an important role in creating and expanding international market relationships that incorporate sustainability and social responsibility. The movement is now compelled to improve opportunities for producers to understand and interface with an expanding web of sustainable value chains as it continues to grow consumer awareness and demand. Fair Trade Town or Fair Trade University initiatives are another avenue of expansion as well as a potential model for effecting norm-changing reforms at higher levels of governance. Farmers, activists, and scholars have all directly or indirectly contributed to this book, and all can and should contribute to addressing the challenges that confront the Fair Trade movement, which must pursue immediate economic strategies as well as longer-term political goals in order to institutionalize the norm change that they have already achieved.

Notes

INTRODUCTION

1. "Fair Trade" is the term used in Australia, Canada, New Zealand, and the United States; "Fairtrade" is the preferred term in the United Kingdom.
2. Retailers that sell Fair Trade products do not have to be certified themselves, although importers and distributors do.
3. Fairtrade International (www.fairtrade.net) and FLO-CERT GmbH (www.flo-cert.net/flo-cert) are the best sources of up-to-date information about Fair Trade standards, products, producers, and traders. Many of the country-specific Fair Trade organizations maintain excellent websites, most notably Fair Trade USA (www.fairtradeusa.org). Because market conditions change, new products are being certified, and Fair Trade is growing and maturing, things can change quickly.
4. The full set of Fairtrade International's generic standards for producers and traders can be found at www.fairtrade.net/fileadmin/user_upload/content/GTS_Feb09_EN.pdf.
5. Claude Williams, African Terroir Winery, interview by the author, July 21, 2006.
6. The critiques mostly aim to inform a strengthened Fair Trade movement by suggesting ways to make the system more fair for producers, both by diminishing knowledge and power differentials between producers and buyers and by transferring more income directly to farmers (see, e.g., Bacon et al. 2008; Jaffe 2007; Renard 2003, 2005).
7. Gavin Fridell (2007, 83-89) describes three overlapping contemporary visions of Fair Trade. The "shaped advantage" perspective is a nongovernmental approach that

focuses on mitigating the negative impacts of neoliberal globalization. The "alternative globalization" approach sees Fair Trade as part of a larger political project. And the "decommodification" approach aims to challenge the values of capitalism.

Chapter 1 FAIR TRADE FROM THE GROUND UP

1 An interactive list from the Fair Trade certifier FLO-Cert GmbH is available here: www.flo-cert.net/flo-cert/main.php?id=10.
2 In 1998 the WTO ruled in favor of the United States and several Latin American governments (acting at the request of Chiquita) to end protected markets for Windward Islands banana farmers (Moberg 2005, 9).
3 This was, at least in part, due to ambiguity built into the tea auction system. All tea is sold at auction, and farmers had no advance knowledge of how much would sell under Fair Trade terms. They only learned this when premium funds arrived in their group's bank accounts. Fairtrade International has since refined the Fair Trade tea protocols to require more transparency.
4 Hired help is not always necessary. For example, a study of cooperatives throughout Africa documents incidents of strong, knowledgeable primary-group leadership (Satgar and Williams 2008). And Josefina Aranda and Carmen Morales (2002) noticed that the Mexican union CEPCO has member-leaders who demonstrate a sophisticated understanding of Fair Trade as a system of buying, price premiums, and the social premium. They gained this understanding not simply by assuming leadership roles but also through their participation in educational and exchange tours to the United States, Canada, and Europe.
5 Importers and sellers of Fair Trade products also pay fees.
6 In 2010 the Fairtrade Labelling Organizations International (FLO) revised the Darjeeling tea standards thus: "Infrastructure projects of benefit to both the plantation management and the workers' community may be approved by FLO if it can be demonstrated to FLO's complete satisfaction that this has the clear approval of the workers. For such projects premium money may be used if the company provides at least 50 percent of the total costs from their own resources as matching funds. In cases where the company lacks immediate capital for such investments, the Joint Body may make a loan to the company to cover the company's 50 percent investment which is to be reimbursed over a period stipulated by the Joint Body at an interest rate which covers at least the rate of inflation over the defined period" (FLO 2010b).
7 Exceptions: In 2005 Nestle introduced a Fair Trade instant coffee in the United Kingdom, and Sam's Club launched a Brazilian Fair Trade–certified "gourmet coffee" in the United States.
8 Alternatively, Marie-Christine Renard (2005) argues that quality conventions are socially constructed to maintain the powerful positions held by buyers in the global North.
9 Biodynamic farming considers the soil and the farm as living organisms. "It

regards maintenance and furtherance of soil life as a basic necessity if the soil is to be preserved for generations, and it regards the farm as being true to its essential nature if it can be conceived of as a kind of individual entity in itself" (Biodynamic Farming and Gardening Association 2008).

10 It is quite likely that US-based Kraft foods will buy Cadbury. Kraft has promised to uphold Cadbury's current commitment to Fair Trade but not necessarily to continue Cadbury's ongoing talks to expand its use of Fair Trade cocoa beans into other brands (Carrell 2010).

11 Because of the necessity for bananas to reach their markets quickly, TransFair USA partnered with Chiquita to import some of its Fair Trade bananas. Given Chiquita's history as the heir of the United Fruit Company and a symbol of US neocolonialism, some Fair Trade activists were outraged. Frundt, in "Fair Bananas" (2009), and Raynolds, in "Fair Trade Bananas" (2007), thoroughly document TransFair's banana campaign and its fallout.

12 A notable exception is the Fair Trade Research Group, whose studies *are* all organized around the same questions and have been well summarized by Murray, Raynolds, and Taylor (2003).

Chapter 2 FAIR TRADE COFFEE IN GUATEMALA

1 Benoit Daviron and Stefano Ponte (2005) and April Linton (2005/2006) describe other efforts to promote sustainability in the coffee industry.

2 Researchers Rebecca Jo Sanborn and Owen Ozier conducted a survey in January and February 2008 that colleagues and I designed as a pilot for an ongoing UC San Diego and UC Berkeley project. Sanborn translated the qualitative responses from the Spanish.

3 Two Guatemalan Fair Trade groups produce honey and were not part of this study.

4 This number includes two that are part of second-tier organizations and not individually registered with FLO-CERT: Association de Desarollo Integral Sostenible de Quetzaltepeque (ADISQUE), which sells through Federación de Cooperativas Agrícolas de Productores de Café de Guatemala (FEDECOCAGUA), and Asociación Integral Unidos para Vivir Mejor (ASUVIM) which sells through Manos Campesinas. Four cooperatives were not surveyed because of travel constraints or safety concerns.

5 There are no explicit rules about how producer groups should spend their accumulated Fair Trade premiums, although of late Fairtrade International requires groups to present, in advance, a plan for approval.

6 Where $p < = 0.05$ or lower.

7 Three cooperative leaders and one union leader did not answer the question. In one case, the interviewee had been newly hired to manage a cooperative's books and had little knowledge of the group's decision-making processes.

8 The number of union leaders surveyed is small because we prioritized reaching cooperative leaders and member-farmers.

9 Farmer, interview by author, February 5, 2008.

10 Producer group leaders actually overestimated side-selling, estimating that members were selling 45 percent of their coffee outside of their cooperative. The question put to leaders asked specifically about *coyotes*, not about side-selling in general.

11 Karen Cebreros, Elan Organic Coffees, interview, August 26, 2008.

Chapter 3 HOW DO PRODUCERS SPEND THE SOCIAL PREMIUM?

1 We use the term "producer group" to encompass cooperatives, unions of cooperatives, and plantations.

2 See Philip Keefer and Stuti Khemani (2005) and Jeffrey Herbst (2000) for discussions of why it is so difficult for fledgling democracies to provide public goods that need to be maintained, such as law and order, defense, contract enforcement, infrastructure, and health and education services.

3 Alfredo Asencio, APECAFÉ, interview by author, July 14, 2003.

4 In fact, most (if not all) research about how consumers respond to Fair Trade or related social-labeling initiatives focuses on benefits to smallholder farmers (Basu and Hicks 2008) or workers (Hiscox and Smyth 2005, Prasad et al. 2004), not communities at large. Even Fairtrade International's Impact Reports do not usually have premium allocation data; they simply state whether or not it is being used "appropriately." An effort is currently underway to standardize these reports and make them more useful to those interested in assessing Fair Trade's social and economic impacts in concrete terms.

5 Telephone interview by author, March 18, 2008.

6 The HDI's standardization and near-universality are not trivial—more nuanced indices are fantastic, but global data conforming to them do not exist.

7 To standardize production across commodities we created three categories by commodity/unit: less than the 33rd percentile, 33rd-66th percentile, and greater than the 66th percentile.

8 The figures are graphic versions of logistic regression analyses. Only variables that are significant at the 0.10 level or lower are included.

9 Although there clearly is some competition between spending on members-only and public health initiatives, nine producer groups did decide to invest in both private and public health projects, e.g., subsidizing members' health care *and* helping build a clinic that anyone may visit.

10 In Central America, a quintal is equal to 46 kilograms or 101.2 pounds. At the time of Utting-Chamorro's study, the Fair Trade premium for coffee was five cents a pound.

Chapter 4 SELLING AND BUYING FAIR TRADE

1 Table 3.1 lists all Fair Trade labelers.

2 Daniel Jaffee (2007, chapter 7) offers a description and critical take on this contract.

3 "The Institute for Marketecology (IMO) Social & Fair Trade Certification ensures that human rights are guaranteed at any stage of production, workers enjoy good and fair working conditions and that smallholder farmers receive a fair share" (Whole Foods Market 2009b).
4 Jessica Hasslocher Johnson, Whole Foods, telephone interview by author, August 31, 2009.
5 I augment Grodnik and Conroy's account with firsthand information from Rick Peyser, Green Mountain's director of social advocacy and coffee community outreach.
6 Rick Peyser, Green Mountain Coffee Roasters, telephone interview by author, August 31, 2009.
7 Ibid.
8 Sandy Yusen, Green Mountain Coffee Roasters, telephone interview by author, August 25, 2009.
9 Ibid.
10 Bezençon and Blili modified a scale developed by Will Low and Eileen Davenport (2005).
11 The following chapter takes a closer look at similar organizations in the United States and the people behind them.
12 The term "greenwash" emerged in the 1970s in reference to the "green" advertising practices of companies with poor environmental records. "Bluewash" stems from the color of the United Nations flag. It refers to similar strategies used to mitigate poor reputations in the area of social justice and human rights (Talbot 2004, 209). In common parlance, activists often use "greenwash" when referring to token placement of Fair Trade products or to other, less rigorous, certifications that are said to address labor and environmental concerns.
13 I return to this topic in chapter 7.
14 In Germany, where, compared to the United Kingdom, Fair Trade is less mainstreamed, consumers with a taste for ethics usually follow the lead of Gepa (a Fair Trade company) and TransFair (now FairTrade Germany), whereas in Britain "both the committed and mainstream [taste for ethics and ethical taste] consumers tended to customize Fair Trade discourses—affirming the sovereignty of the customer" (Varul and Wilson-Kovacs 2008, 14).
15 See, e.g., Cathy Cobb-Walgren et al. (1995), Deirdre Shaw and Ian Clarke (1999).
16 See Michael Hiscox and Nicholas Smythe (2006) for an excellent review.
17 Becchetti and Rosati's pool of interviewees was fairly diverse: 32 percent students, 26 percent retired people, 12 percent housewives and 8 percent professionals. On average, interviewees spent €20 per month on Fair Trade products, a figure representing 0.6 to 1.3 percent of their household incomes. Fair Trade consumption did not rise with income.
18 Of course, Becchetti and Rosati's study was limited to people whose commitment to ethical consumption was at least high enough to propel them through the door of a World Shop. More than half of their interviewees were aware of the fact that Fair Trade products are also sold in mainstream outlets such as major supermarket

chains, but only 17 percent said they would shop in these places. However, 50 percent said they would do so if mainstream distributors were to adopt more ethical business practices overall.

19 De Pelsmacker and Janssens also found, not surprisingly, that skepticism about the power of consumer-based efforts to allay global inequalities is negatively related to Fair Trade.

20 Even in this sample, 43 percent said they had never purchased a Fair Trade product.

21 Only respondents who claimed to buy coffee for home consumption were asked this question.

22 The multivariate analysis conducted by Hertel, Scruggs, and Heidkamp (2009) does not support the connection between thinking that a minimum standard of living is a human right and willingness to pay more for a Fair Trade coffee. This is probably because this analysis controlled for income and education. Some respondents who were *willing* to pay are not actually *able* to pay.

23 E-mail correspondence with Shareen Hertel and research assistant Chris Jeffords, August 20, 2010.

Chapter 5 FAIR TRADE ACTIVISTS IN THE UNITED STATES

1 This chapter is based on part of Rebecca Kahn's master's thesis (2009). The larger research project included fieldwork in an Ethiopian coffee-producing community and interviews with people who consume Fair Trade products but do not actively promote them.

2 Alexandra Mello, Northampton Fair Trade Partnerships, interview by Rebecca Kahn, October 17, 2008.

3 Barth Anderson, Barrington Roasting Company, interview by Rebecca Kahn, October 14, 2008.

4 Taylor Mork, Crop to Cup Coffee, interview by Rebecca Kahn, October 28, 2008.

5 Judith Belasco, Hazon, interview by Rebecca Kahn, October 13, 2008.

6 Pattie Cippi Harte, Jewish Community Centers Association, interview by Rebecca Kahn, October 10, 2008.

7 We do not encourage readers to assume that "local is better" under all circumstances. See, e.g., Melanie DuPuis and David Goodman (2005) and Peter Singer and Jim Mason (2007).

8 Yuri Friman, Amherst Fair Trade Partnership, interview by Rebecca Kahn, October 15, 2008.

9 Alexandra Mello, Northampton Fair Trade Partnerships, interview by Rebecca Kahn, October 17, 2008.

10 Stephanie Celt, Washington Fair Trade Coalition, interview by Rebecca Kahn, November 19, 2008.

11 Ron Zisa, Park Slope Food Coop, interview by Rebecca Kahn, October 7, 2008.

12 Barth Anderson, Barrington Roasting Company, interview by Rebecca Kahn, October 14, 2008.

13 Ron Zisa, Park Slope Food Coop, interview by Rebecca Kahn, October 7, 2008.

Chapter 6 A FAIR TRADE UNIVERSITY

1 The first was University of Wisconsin, Oshkosh. Wisconsin students looked to the United Kingdom's Fairtrade Foundation for guidance, whereas UCSD worked with Fair Trade USA. UCSD's policy is somewhat stricter in that it encompasses more products and stakeholders.
2 Viraf Soroushian, e-mail correspondence, September 30, 2009.
3 In May and June of 2009, I interviewed nearly all of the former and current students, staff members, and administrators who are or have been involved in the campaign or the advisory committee that stemmed from it—twenty-four people in total. My interviewees were very generous with their time, in part because they wish for their experiences to inform the work of Fair Trade activists and sustainability-minded leaders on other campuses.
4 Documented by TransFair USA. http://getinvolved.transfairusa.org/site/News2?page=NewsArticle&id=5161&news_iv_ctrl=1061
5 Documented by TransFair USA. http://getinvolved.transfairusa.org/site/News2?page=NewsArticle&id=5151&news_iv_ctrl=1061
6 Claire Tindula, UCSD, interview by author, June 11, 2009.
7 Subsequently, Linneman transferred the group's 501(c)(3) status to another San Diego nonprofit.
8 At that time, Associated Students operated several businesses on campus.
9 The committee reports to vice-chancellors Steve Relyea, external and business affairs, and Penny Rue, student affairs.
10 June Reyes, UCSD, interview by author, June 12, 2009.
11 Fran Avendaño, UCSD, interview by author, June 12, 2009.
12 Rishi Ghosh, UCSD, interview by author, May 7, 2009.
13 Chris Westling, UCSD, interview by author, May 28, 2009.
14 Yvonne Macon, e-mail correspondence, February 1, 2010.
15 Rishi Ghosh, UCSD, interview by author, May 7, 2009.
16 Boshen Jia, phone interview by author, June 12, 2009.
17 Rishi Ghosh, UCSD, interview by author, May 7, 2009.
18 Maggie Souder, UCSD, interview by author, June 11, 2009.
19 Maggie Souder, UCSD, interview by author, June 11, 2009.
20 Fran Avendaño, UCSD, interview by author, May 21, 2009.
21 Claire Tindula, UCSD, interview by author, June 11, 2009.
22 Jessica Wall, UCSD, interview by author, May 5, 2009.
23 Rishi Ghosh, UCSD, interview by author, May 7, 2009.

Chapter 7 GROWING FAIR TRADE

1 All but one interview in this chapter were conducted in August 2009.
2 A detailed list of criteria for certified factories can be found at www.fairtradecertified.org/sites/default/files/FTC_Apparel-Obligations_of_CMT_Facility.pdf.

3 Heather Franzese, Fair Trade USA, telephone interview conducted by the author, July 22, 2011.
4 Ibid.
5 See, for example, Locke et al. 2006.
6 SA8000, created in 1997 by Social Accountability International, includes standards pertaining to health and safety, working hours, child labor, forced labor, discrimination, freedom of association and collective bargaining, wages, and discipline (SAAS 2011).
7 The "perverse outcomes" Guthman refers to are created scarcity, barriers to entry, "the worst sort of 'green-washing,'" and the fetishization of ethicality itself (2008, 205).
8 The information in this section derives from Fair Trade Resource Network's Webinar #117: A Community Discussion About What Fair Trade USA Leaving FLO Means for Producers and the Movement. Part 1 with Paul Rice. October 12, 2011. Available at www.fairtraderesource.org/learn-up/buy-ftrn-publications/#webinars.

References

Anner, Mark, and Peter Evans. 2004. Building bridges across a double divide: Alliances between US and Latin American labour and NGOs. *Development in Practice* 14 (1/2): 34–47.

Aranda, Josefina, and Carmen Morales. 2002. Poverty alleviation through participation in Fair Trade coffee networks: The case of CEPCO, Oaxaca, Mexico. Center for Fair and Alternative Trade. Accessed September 16, 2011. http://welcome2.libarts.colostate.edu/centers/cfat/wp-content/uploads/2009/09/Case-Study-CEPCO-Oaxaca-Mexico.pdf.

Aparicio, Susana, Sutti Ortiz, and Nidia Tadeo. 2009. Have private supermarket norms benefited labourers? Lemon and sweet citrus production in Argentina. In *The Global Governance of Food*, edited by Sara Curran, April Linton, Abigail Cooke, and Andrew Schrank, 65–78. London: Routledge.

Arnot, Chris, Peter C. Boxall, and Sean B. Cash. 2006. Do ethical consumers care about price? A revealed preferences analysis of Fair Trade coffee purchases. *Canadian Journal of Agricultural Economics* 54: 555–65.

Audebrand, Luc K., and Thierry C. Pauchant. 2009. Can the Fair Trade movement enrich traditional business ethics? An historical study of its founders in Mexico. *Journal of Business Ethics* 87: 343–53.

Bacon, Christopher M., V. Ernesto Méndez, María Eugenia Flores Gómez, Douglas Stuart, and Sandro Raúl Díaz Flores. 2008. Are sustainable coffee certifications enough to secure farmer livelihoods? The millennium development goals and Nicaragua's Fair Trade cooperatives. *Globalizations* 5 (2): 259–74.

Barrientos, Stephanie, and Sally Smith. 2007. Mainstreaming Fair Trade in global production networks: Own brand fruit and chocolate in U.K. supermarkets. In *Fair Trade: The Challenges of Transforming Globalization,* edited by Laura T. Raynolds, Douglas Murray, and John Wilkinson, 103–22. London: Routledge.

Bartley, Tim. 2003. Certifying forests and factories: States, social movements, and the rise of private regulation in the apparel and forest products fields. *Politics & Society* 31 (3): 433–64.

Basu, Arnab K., and Robert L. Hicks. 2008. Label performance and the willingness to pay for Fair Trade coffee: A cross-national perspective. *International Journal of Consumer Studies* 32: 470–78.

Bates, Robert. 1997. *Open-Economy Politics: The Political Economy of the World Coffee Trade.* Princeton, NJ: Princeton University Press.

Becchetti, Leonardo, and Furio Camillo Rosati. 2007. Global social preferences and the demand for socially responsible products: Empirical evidence from a pilot study on Fair Trade consumers. Center for International Studies on Economic Growth, Working Paper 91. Accessed September 16, 2011. http://dspace.uniroma2.it/dspace/bitstream/2108/308/1/SSRN-id962488.pdf.

Beck, Ulrich. 1992. *Risk Society: Towards a New Modernity.* London: Sage.

Besky, Sarah. 2010. Colonial pasts and Fair Trade futures: Changing modes of production and regulation on Darjeeling tea plantations. In *Fair Trade and Social Justice: Global Ethnographies,* edited by Sarah Lyon and Mark Moberg, 97–124. New York: New York University Press.

Bezençon, Valéry, and Sam Blili. 2009. Fair Trade managerial practices: Strategy, organisation, and engagement. *Journal of Business Ethics* 9 (1): 95–113.

Biel, Anders, Daniel Eek, and Tommy Gärling. 1997. Distributive justice and willingness to pay for municipality child care. *Social Justice Research* 10: 63–80.

Biodynamic Farming and Gardening Association. 2008. What is biodynamic agriculture? Accessed August 10, 2011. www.biodynamics.com/biodynamics.html.

Brock, Gillian. 2005. Does obligation diminish with distance? *Ethics Place and Environment* 8 (1): 3–20.

Bryant, Raymond L., and Michael K. Goodman. 2004. Consuming narratives: The political ecology of "alternative" consumption. *Transactions of the Institute of British Geographers* 29: 344–66.

Buchanan, James. 1965. An economic theory of clubs. *Econometrica* 32: 1–14.

Buechler, Steven M. 2000. *Social Movements in Advanced Capitalism.* New York: Oxford University Press.

Cáceres, Aurélie Carimentran, and John Wilkinson. 2007. Fair Trade and quinoa from the southern Bolivian altiplano. In *Fair Trade: The Challenges of Transforming Globalization,* edited by Laura T. Raynolds, Douglas Murray, and John Wilkinson, 180–99. London: Routledge.

Cadbury. 2009. Cadbury dairy milk is going fairtrade. Accessed January 5, 2010. www.cadbury.co.uk/cadburyandchocolate/howchocolateismade/Pages/cdmfairtrade.aspx.

Cailleba, Patrice, and Herbert Casteran. 2010. Do ethical values work? A quantitative study of the impact of Fair Trade coffee on consumer behavior. *Journal of Business Ethics* 97: 613–24.

Carrell, Severin. 2010. Kraft pledges to honor Cadbury's Fairtrade sourcing commitments. *The Guardian,* January 25. Accessed February 17, 2010. www.guardian.co.uk/business/2010/jan/23/kraft-cadbury-fairtrade.

Catholic Relief Services (CRS). 2010. Building solidarity through Fair Trade partners. Accessed August 2, 2011. www.crsfairtrade.org/building-solidarity/.

Chandler, Paul, 2006. Fair trade and global justice. *Globalizations* 3 (2): 255–57.

Clarke, Nick, Clive Barnett, Paul Cloke, and Alice Malpass. 2007. The political rationalities of Fair-Trade consumption in the United Kingdom. *Politics and Society* 35 (4): 583–607.

Cobb-Walgren, Cathy J., Carolyn A. Ruble, and Naveen Donthu. 1995. Brand equity, brand preference, and purchase intent. *Journal of Advertising* 24: 25–41.

Cooke, Abigail, Sara R. Curran, April Linton, and Andrew Schrank. 2008. Conclusion: Negotiating the dynamics of global complexity. *Globalizations* 5 (2): 319–28.

Cooley, Jack P., and Daniel A. Lass. 1998. Consumer benefits from community supported agriculture membership. *Review of Agricultural Economics* 20 (1): 227–37.

Cornes, Richard and Todd Sandler. 1996. *The Theory of Externalities: Public Goods and Club Goods.* New York: Cambridge University Press.

Crossen, Scott, John Orbell, and Holly Arrow. 2004. Social poker: A laboratory test of predictions from Club Theory. *Rationality and Society* 16 (2): 225–248.

Cup of Excellence. 2011. What is Cup of Excellence? Accessed September 16, 2011. www.cupofexcellence.org/WhatisCOE/tabid/184/Default.aspx.

David, Scott M. 2000. *Brand Asset Management: Driving Profitable Growth through Your Brands.* San Francisco: Jossey-Bass.

Davies, Iain A. 2009. Alliances and networks: Creating success in the UK Fair Trade market. *Journal of Business Ethics* 86: 109–26.

Daviron, Benoit, and Stefano Ponte. 2005. *The Coffee Paradox: Global Markets, Commodity Trade and the Elusive Promise of Development.* London: Zed Books.

DeCarlo, Jacqueline. 2007. *Fair Trade: A Beginner's Guide.* Oxford: Oneworld Publications.

DeChaine, D. Robert. 2005. *Global Humanitarianism: NGOs and the Crafting of Community.* London: Lexington Books.

De Cremer, David. 2003. Noneconomic motives predicting cooperation in public good dilemmas: The effect of received respect on contributions. *Social Justice Research* 16 (4): 367–77.

de Pelsmacker, Patrick, and Wim Janssens. 2007. A model for Fair Trade buying behaviour: The role of perceived quantity and quality of information and of product-specific attitudes. *Journal of Business Ethics* 75: 361–80.

Doherty, Bob, and Sophi Tranchell. 2005. New thinking in international trade? A case study of the Day Chocolate Company. *Sustainable Development* 13: 166–76.

Dolan, Catherine S. 2008. In the mists of development: Fairtrade in the Kenyan tea fields. *Globalizations* 5 (2): 305–18.

———. 2010. Fractured ties: The business of development in Kenyan Fair Trade tea. In *Fair Trade and Social Justice: Global Ethnographies*, edited by Sarah Lyon and Mark Moberg, 145–57. New York: New York University Press.

Doran, Carolyn Josephine. 2009. The role of personal values in Fair Trade consumption. *Journal of Business Ethics* 84: 549–63.

DuPuis, E. Melanie, and Sean Gillon. 2009. Alternative modes of governance: Organic as civic engagement. *Agriculture and Human Values* 26 (1–2): 43–56.

DuPuis, E. Melanie, and David Goodman. 2005. Should we go "home" to eat? Towards a reflexive politics of localism. *Journal of Rural Studies* 21 (3): 359–71.

Du Toit, Andries. 2002. Globalizing ethics: Social technologies of private regulation and the South African wine industry. *Journal of Agrarian Change* 2 (3): 356–80.

Eek, Daniel, and Anders Biel. 2003. The interplay between greed, efficiency, and fairness in public-goods dilemmas. *Social Justice Research* 16 (3): 195–215.

Eek, Daniel, Anders Biel, and Tommy Gärling. 1998. The effect of distributive justice on willingness to pay for municipality child care: An extension of the GEF hypothesis. *Social Justice Research* 11: 121–42.

Equal Exchange. 2008. Our story. Accessed August 24, 2009. www.equalexchange.coop/story.

Escobar, Arturo. 2001. Culture sits in places: Reflections on globalism and subaltern strategies of localization. *Political Geography* 20: 139–74.

Fair Trade Deutschland. 2011. Faire Woche 2009. Accessed July 3, 2011. www.fairtrade-deutschland.de.

Fairtrade Foundation. 2009a. Fairtrade school identity manual. Accessed September 16, 2011. www.fairtrade.org.uk/includes/documents/cm_docs/2010/f/1_fairtrade_foundation_school_identity_manual_december_2009.pdf.

———. 2009b. Fairtrade fortnight. Accessed July 3, 2011. www.fairtrade.org.uk/get_involved/fairtrade_fortnight/fairtrade_fortnight_2009/default.aspx.

———. 2009c. About Fairtrade universities and colleges. Accessed February 5, 2010. www.fairtrade.org.uk/get_involved/campaigns/fairtrade_universities/about_fairtrade_universities.aspx.

Fair Trade Towns USA. 2010. Fair Trade towns toolkit. Accessed September 16, 2011. www.fairtradetownsusa.org/wp-content/uploads/downloads/2010/11/ftt_toolkit_final.pdf.

———. 2011. Towns. Accessed September 16, 2011. www.fairtradetownsusa.org/towns.

Fair Trade USA. 2010. Apparel and linens program. Accessed August 1, 2011. www.fairtradecertified.org/certification/producers/apparel-linens.

———. 2011a. Who we are. Accessed August 2, 2011. www.transfairusa.org/about-fair-trade-usa/who-we-are.

———. 2011b. ASOPROBAN. Accessed September 1, 2011. www.transfairusa.org/node/367.

———. 2011c. 2010 Almanac. Accessed September 1, 2011. www.transfairusa.org/sites/default/files/Almanac%202010_0.pdf.

Feagan, Robert. 2007. The place of food: Mapping out the "local" in local food systems. *Progress in Human Geography* 31 (1): 23–42.

Featherstone, Mike. 1991. *Consumer Culture and Postmodernism*. London: SAGE.
FLO (Fairtrade International). 2004. Shopping for a better world. Annual report 03/04. Accessed September 16, 2011. www.fairtrade.net/fileadmin/user_upload/content/AR_03-04_screen_final-1.pdf.
———. 2006. Standard operating procedure ("SOP") (summary) development of Fairtrade standards. Accessed September 16, 2011. www.fairtrade.net/fileadmin/user_upload/content/SOP_Public_Development_of_Fairtrade_Standards.pdf.
———. 2008. An inspiration for change. Annual report 2007. Accessed September 16, 2011. www.fairtrade.net/fileadmin/user_upload/content/FLO_AR2007_low_res.pdf.
———. 2010a. Growing stronger together (annual report 2009–10). Accessed September 16, 2011. www.fairtrade.net/fileadmin/user_upload/content/2009/resources/FLO_Annual-Report-2009_komplett_double_web.pdf.
———. 2010b. Fairtrade standards for tea for hired labour. Accessed September 16, 2011. www.fairtrade.net/fileadmin/user_upload/content/2009/standards/documents/2010-12-22_EN_Tea_HL_2.pdf.
———. 2011a. Fairtrade minimum price and Fairtrade premium table. Accessed September 16, 2011. www.fairtrade.net/fileadmin/user_upload/content/2009/standards/documents/2011-08-15_EN_Fairtrade_Minimum_Price_and_Premium_Table.pdf.
———. 2011b. Q & A on Fairtrade International and Fair Trade USA. http://fairtrade.net/897.0.html. Accessed October 17, 2011.
Freidberg, Susanna. 2004. *French Beans and Food Scares: Culture and Commerce in an Anxious Age*. Oxford: Oxford University Press.
Fridell, Gavin. 2007. *Fair Trade Coffee: The Prospects and Pitfalls of Market-Driven Social Justice*. Toronto: University of Toronto Press.
———. 2009. The co-operative and the corporation: Competing visions of the future of Fair Trade. *Journal of Business Ethics* 86: 81–95.
Frundt, Henry J. 2009. *Fair Bananas: Farmers, Workers, and Communities Strive to Change an Industry*. Tucson: University of Arizona Press.
Fung, Archon, Dara O'Rourke, and Charles Sabel. 2001. *Can We Put an End to Sweatshops?* Boston: Beacon Press.
Garza, Víctor Pérezgrovas, and Edith Cervantes Trejo. 2002. Poverty alleviation through participation in Fair Trade coffee networks: The case of Unión Majomut, Chiapas, Mexico. Center for Fair and Alternative Trade. Accessed October 5, 2004. www.cfat.colostate.edu/research/.
Gendron, Corinne, Véronique Bisaillon, and Ana Isabel Otero Rance. 2009. The institutionalization of Fair Trade: More than just a degraded form of social action. *Journal of Business Ethics* 86:63–79.
Gereffi, Gary, John Humphrey, and Timothy Sturgeon. 2006. The governance of global value chains. *Review of International Political Economy* 12 (1): 78–104.
Getz, Christy, and Aimee Shreck. 2006. What organic and Fair Trade labels do not tell us: Towards a place-based understanding of certification. *International Journal of Consumer Studies* 30 (5): 490–501.
Gibson-Graham, J. K. 2005. Surplus possibilities: Postdevelopment and community economics. *Singapore Journal of Tropical Geography* 26 (1): 4–26.

Global Exchange. 2007. There's nothing magical about child labor! Make my Wonka Bar Fair Trade! Accessed January 6, 2009. www.globalexchange.org/campaigns/fairtrade/cocoa/charlieaction.html.

———. 2010. Fourth annual reverse trick-or-treating! Accessed October 20, 2010. www.globalexchange.org/campaigns/fairtrade/cocoa/reversetrickortreating/reversetrickortreating.html.

González Cabañas, Alma Amalia. 2002. Evaluation of the current and potential poverty alleviation benefits of participation in the Fair Trade market: The case of Unión La Selva, Chiapas, Mexico. Center for Fair and Alternative Trade. Accessed October 5, 2004. www.cfat.colostate.edu/research/.

Goodman, David, and E. Melanie DuPuis. 2002. Knowing food and growing food: Beyond the production-consumption debate in the sociology of agriculture. *Sociologia Ruralis* 42 (1): 5–22.

Goodman, Michael K. 2004. Reading Fair Trade: Political ecological imaginary and the moral economy of Fair Trade foods. *Political Geography* 23: 891–915.

Goul Andersen, Jørgen, and Mette Tobiasen. 2006. Who are these political consumers anyway? Survey evidence from Denmark. In *Politics, Products and Markets: Exploring Political Consumerism Past and Present*, edited by Michele Micheletti, Andreas Follesdal, and Dietlind Stolle, 203–21. London: Transaction Publishers.

GourmetSpot. 2009. Chocolate fact sheet. Accessed July 15, 2010. www.gourmetspot.com/factschocolate.htm.

Gressler, Charis, and Sophia Tickell. 2002. *Mugged: Poverty in Your Coffee Cup*. Oxfam International.

Grodnik, Ann, and Michael E. Conroy. 2007. Fair Trade coffee in the United States: Why companies join the movement. In *Fair Trade: The Challenges of Transforming Globalization,* edited by Laura Raynolds, Douglas Murphy, and John Wilkinson, 83–102. London: Routledge.

Guthman, Julie. 2008. Unveiling the unveiling: Commodity chains, fetishism, and the "value" of voluntary, ethical food labels. In *Frontiers of Commodity Chain Research*, edited by Jennifer Bair, 190–206. Stanford, CA: Stanford University Press.

Herbst, Jeffrey. 2000. *States and Power in Africa*. Princeton, NJ: Princeton University Press.

Hertel, Shareen. 2006. *Unexpected Power: Conflict and Change among Transnational Activists*. Ithaca, NY: Cornell University Press.

Hertel, Shareen, Lyle Scruggs, and C. Patrick Heidkamp. 2009. Human rights and public opinion: From attitudes to actions. *Political Science Quarterly* 124 (3): 443–59.

Hiscox, Michael, and Nicholas Smythe. 2006. Is there consumer demand for improved labor standards? Evidence from field experiments in social labeling. http://dev.wcfia.harvard.edu/sites/default/files/HiscoxSmythND.pdf.

Hohenegger, Beatrice. 2006. *Liquid Jade: The Story of Tea from East to West*. New York: St. Martin's Press.

Hunt, Shelby D., and Scott Vitell. 1993. The general theory of marketing ethics: A retrospective and revision. In *Ethics in Marketing*, edited by N. Craig Smith and John A. Quelch, 775–64. Homewood, IL: Irwin.

Jaffee, Daniel. 2007. *Brewing Justice: Fair Trade Coffee, Sustainability, and Survival.* Berkeley: University of California Press.

Jaffee, Daniel, Jack R. Kloppenburg Jr., and Mario B. Monroy. 2004. Bringing the "moral change" home: Fair Trade within the North and within the South. *Rural Sociology* 69 (2): 169–96.

Kahn, Rebecca. 2009. Voices of Fair Trade coffee consumption in the United States. Master's thesis, University of Auckland.

Kaul, Inge, Pedro Conceição, Katell Le Goulven, and Ronald U. Mendoza. 2003. How to improve the provision of global public goods. In *Providing Global Public Goods: Managing Globalization* edited by Inge Kaul, Pedro Conceição, Katell Le Goulven, and Ronald U. Mendoza, 21–58. New York and Oxford: Oxford University Press.

Keck, Margaret, and Katherine Sikkink. 1998. *Activists Beyond Borders: Advocacy Networks in Transnational Politics.* Ithaca, NY: Cornell University Press.

Keefer, Philip, and Stuti Khemani. 2005. Democracy, public expenditures, and the poor: Understanding political incentives for providing public services. *World Bank Research Observer* 20 (1): 1–27.

Kleine, Dorothea. 2008. Negotiating partnerships, understanding power, doing action research on Chilean Fairtrade wine value chains. *The Geographical Journal* 174 (2): 109–23.

KPMG. 2009. Sustainability, corporate social responsibility through an audit committee lens. Accessed January 3, 2010. www.kpmg.com.

Kraft Foods. 2010. Responsibility. Accessed May 2, 2010. www.kraftfoodscompany.com/Responsibility/index.aspx.

Kristoff, Nicholas, and Cheryl WuDunn. 2009. *Half the Sky: Turning Oppression into Opportunity for Women Worldwide.* New York: Knopf.

Kruger, Sandra, and Andries Du Toit. 2007. Reconstruction fairness: Fair Trade conventions and worker empowerment in South African horticulture. In *Fair Trade: The Challenges of Transforming Globalization,* edited by Laura T. Raynolds, Douglas Murphy, and John Wilkinson, 200–219. London: Routledge.

Lacy, William B. 2000. Empowering communities through public work, science, and local food systems: Revisiting democracy and globalization. *Rural Sociology* 65 (1): 3–26.

Le Velly, Roman. 2006. Le commerce équitable: Des échanges marchands contre et dans le marché. *Revue français de sociologie* 47 (2): 319–40.

Levi, Margaret, and April Linton. 2003. Fair Trade: A cup at a time? *Politics and Society* 31 (3): 407–32.

Lewis, Jessa, and David Runsten. 2008. Is Fair Trade–organic coffee sustainable in the face of migration? Evidence from a Oaxacan community. *Globalizations* 5 (2): 275–90.

Linton, April. 2005. Partnering for sustainability: Business-NGO alliances in the coffee industry. *Development in Practice* 15 (3/4): 600–614.

———. 2008. A niche for sustainability: Fair labor and environmentally sound practices in the specialty coffee industry. *Globalizations* 5 (2): 231–45.

Linton, April, Cindy Chiayuan Liou, and Kelly Anne Shaw. 2004. A taste of trade justice: Marketing global social responsibility via Fair Trade coffee. *Globalizations* 1 (2): 223–46.

Locke, Richard, Fei Qin, and Alberto Brause. 2006. Does monitoring improve labor standards? Lessons from Nike. MIT Sloan School of Management, Working Paper 4612-06. Accessed October 4, 2007. www.wcfia.harvard.edu/sites/default/files/privategovernance_ws_lockeromis.pdf.

Low, Will, and Eileen Davenport. 2005. Has the medium (roast) become the message? The ethics of marketing Fair Trade in the mainstream. *International Marketing Review* 22 (5): 494–511.

———. 2009. Organizational leadership, ethics and the challenges of marketing fair and ethical trade. *Journal of Business Ethics* 86: 97–108.

Luna, Kira. 2008. Selling on the side: A look at selling practices within Guatemalan Fair Trade coffee cooperatives. Student research paper mentored by April Linton, University of California, San Diego.

Lyon, Sarah. 2002. Evaluation of the actual and potential benefits for the alleviation of poverty through the participation in Fair Trade networks: Guatemalan case study. Center for Fair and Alternative Trade. Accessed February 25, 2003. www.cfat.colostate.edu/research.

———. 2006. Evaluating Fair Trade consumption: Politics, defetishization, and producer participation. *International Journal of Consumer Studies* 30 (5): 452–64.

———. 2007. Fair Trade coffee and human rights in Guatemala. *Journal of Consumer Policy* 30 (3): 241–61.

———. 2010. A market of our own: Women's livelihoods and Fair Trade markets. In *Fair Trade and Social Justice: Global Ethnographies*, edited by Sarah Lyon and Mark Moberg, 125–46. New York: New York University Press.

Macchiette, Bart, and Abhijit Roy. 1991. Direct marketing to the credit card industry: Utilizing the concept of affinity. *Journal of Direct Marketing* 5 (2): 34–43.

Maggie's Functional Organics. 2011. Maggie's apparel: Now Fair Trade certified. Accessed August 13, 2011. www.maggiesorganics.com/2010_howitsmade.php.

Mansvelt, Juliana. 2007. Geographies of consumption: Citizenship, space, and practice. *Progress in Human Geography*, 1–13.

Massey, Douglas, Joaquin Arango, Graeme Hugo, Ali Kouaouci, Adela Pellegrino, and J. Edward Taylor. 1993. Theories of international migration: A review and appraisal. *Population and Development Review* 19 (30): 431–66.

Max Havelaar France. 2009. Tous mobilisés pour la Quinzaine du commerce équitable. Accessed November 30, 2009. www.maxhavelaarfrance.com.

McEvedy, Allegra. 2009. Worthy, but tasty. *The Guardian*, February 25. www.guardian.co.uk/lifeandstyle/2009/feb/25/wine-fairtrade.

McGregor, Andrew. 2007. Development, foreign aid and post-development in Timor-Leste. *Third World Quarterly* 28 (1): 155–70.

McKinnon, Katharine. 2007. Postdevelopment, professionalism, and the politics of participation. *Annals of the Association of American Geographers* 97 (4): 772–85.

Méndez, V. Ernesto. 2002. Fair Trade networks in two coffee cooperatives of western El

Salvador: An analysis of insertion through a second-level organization. Center for Fair and Alternative Trade. Accessed July 20, 2003. www.cfat.colostate.edu/research/.

Micheletti, Michele. 2003. *Politics and Shopping*. New York: Palgrave.

Milinski, Manfred, Dirk Semmann, and Hans-Jürgen Krambeck. 2002. Reputation helps solve the "tragedy of the commons." *Nature* 415: 424–26.

Moberg, Mark. 2005. Fair Trade and eastern Caribbean banana farmers: Rhetoric and reality in the anti-globalization movement. *Human Organization* 64 (1): 4–15.

———. 2010. A new world? Neoliberalism and Fair Trade farming in the Eastern Caribbean. In *Fair Trade and Social Justice: Global Ethnographies*, edited by Sarah Lyon and Mark Moberg, 47–71. New York: New York University Press.

Mohan, Sushil. 2010. *Fair Trade without the Froth*. Hobart Paper 170. London: The Institute for Economic Affairs.

Moseley, William G. 2007. Transformation and the delinquent South African wine connoisseur. *Cape Argus*, March 19, p. 13.

———. 2008. Fair Trade wine: South Africa's post-apartheid vineyards and the global economy. *Globalizations* 5 (2): 291–304.

Murray, Douglas, Laura T. Raynolds, and Peter Leigh Taylor. 2003. One cup at a time: Fair Trade and poverty alleviation in Latin America. Center for Fair and Alternative Trade. Accessed March 25, 2004. www.cfat.colostate.edu/research.

Mutersbaugh, Tad. 2002. Migration, common property, and communal labor: Cultural politics and agency in a Mexican village. *Political Geography* 21: 473–94.

Nicholls, Alex, and Charlotte Opal. 2005. *Fair Trade: Market-Driven Ethical Consumption*. London: Sage.

Ozcaglar-Toulouse, Nil, Edward Shiu, and Deirdre Shaw. 2006. In search of Fair Trade: Ethical consumer decision making in France. *International Journal of Consumer Studies* 30 (5): 502–15.

Parks, Craig D., Lawrence J. Sanna, and Susan R. Berel. 2001. Actions of similar others and inducements to cooperate in social dilemmas. *Perspectives on Social Psychology Bulletin* 27: 345–54.

Pattie, Charles, Patrick Seyd, and Paul Whitely. 2003. Citizenship and civic engagement: Attitudes and behaviour in Britain. *Political Studies* 51: 443–68.

Pauwels, Luc. 2000. Taking the visual turn in research and scholarly communications: Key issues in developing a more visually literate (social) science. *Visual Studies* 15 (1): 7–14.

Pendergrast, Mark. 1999. *Uncommon Grounds: The History of Coffee and How It Transformed Our World*. New York: Basic Books.

Pirotte, Gautier, Geoffrey Pleyers, and Marc Poncelet. 2006. Fair Trade coffee in Nicaragua and Tanzania: A comparison. *Development in Practice* 16 (5): 441–51.

prAna 2011. Sustainable products. Accessed August 11, 2011. www.prana.com/sustainability/products.

Pringle, Hamish, and Marjorie Thompson. 1999. *Brand Spirit: How Cause-Related Marketing Builds Brands*. Chicester, UK: John Wiley & Sons.

Raynolds, Laura T. 2002. Consumer/producer links in Fair Trade coffee networks. *Sociological Ruralis* 42: 404.

———. 2007. Fair Trade bananas: Broadening the movement and the market in the United States. In *Fair Trade: The Challenges of Transforming Globalization*, edited by Laura T. Raynolds, Douglas Murphy, and John Wilkinson, 63–82. London: Routledge.

———. 2008. Mainstreaming Fair Trade coffee: From partnership to traceability. *World Development* 37 (6): 1083–93.

Raynolds, Laura T., and Siphelo Unathi Ngcwangu. 2010. Fair Trade rooibus tea: Connecting South African producers and American consumer markets. *Geoforum* 41:74–83.

Renard, Marie-Christine. 2003. Fair Trade: Quality, market, and conventions. *Journal of Rural Studies* 19: 87–96.

———. 2005. Quality certification, regulation, and power in Fair Trade. *Journal of Rural Studies* 21: 419–31.

Ronchi, Loraine. 2002. The impact of Fair Trade on producers and their organisations: A case study with Coocafé in Costa Rica. Working Paper No. 11, Poverty Research Unit at Sussex, University of Sussex, Falmer, Brighton, UK.

SAAS (Social Accountability Accreditation Services). 2011. SA8000 standard. Accessed September 17, 2011. www.saasaccreditation.org/certSA8000.htm.

Sanborn, Rebecca Jo. 2008. Fairtrade coffee production in Guatemala. Report produced for the University of California, San Diego. Available from the author.

Santos, Boaventura de Sousa. 2003. The World Social Forum: Toward a counter-hegemonic globalisation. *XXIV International Congress of the Latin American Studies Association*. Dallas: World Social Forum.

Satgar, Vishwas, and Michelle Williams. 2008. The passion of the people: Successful cooperative experiences in Africa. Cooperative and Policy Alternative Center (COPAC). Accessed October 15, 2010. www.copac.org.za/publications/passion-people-successful-cooperative-experiences-africa.

Schoenholt, David. 2001. The Fair Trade ideal: The ultimate answer for sustainability? *Tea and Coffee* 175 (11). Accessed September 16, 2011. www.teaandcoffee.net/1101/special.htm.

Shreck, Aimee. 2005. Resistance, redistribution, and power in the Fair Trade banana initiative. *Agriculture and Human Values* 22: 17–29.

Seidman, Gay W. 2007. *Beyond the Boycott: Labor Rights, Human Rights, and Transnational Activism.* New York: Russell Sage Foundation.

Sen, Amartya. 1999. *Development as Freedom*. New York: Anchor Books.

Shaw, Deirdre, and Ian Clarke. 1999. Belief formation in ethical consumer groups: An exploratory study. *Marketing Intelligence and Planning* 17 (2): 109–19.

Shaw, Deirdre, Terry Newholm, and Roger Dickinson. 2006. Consumption as voting: An exploration of consumer empowerment. *European Journal of Marketing* 40 (9/10): 1049–67.

Shaw, Deirdre, and Edward Shiu. 2003. Ethics in consumer choice: A multivariate modelling approach. *European Journal of Marketing* 37 (10): 1485–98.

Shuman, Michael, Alissa Barron, and Wendy Wasserman. 2009. Community food enterprise: Local success in a global marketplace. Wallace Center at Winrock International.

Accessed November 5, 2009. www.communityfoodenterprise.org/book-pdfs/CFE%20-%20okuapa-kokoo_view.pdf.

Sick, Deborah. 2008. Coffee farming, families and Fair Trade in Costa Rica: New markets, same old problems? Latin American Research Review 43 (3): 193–208.

Simpson, Charles R., and Anita Rapone. 2000. Community development from the ground up: Social justice coffee. *Human Ecology Review* 7 (1): 46–57.

Singer, Peter, and Jim Mason. 2007. *The Ethics of What We Eat: Why Our Food Choices Matter*. Emmaus, PA: Rodale.

Smith Maguire, Jennifer, and Kim Stanway. 2008. Looking good: Consumption and the problems of self-production. *European Journal of Cultural Studies* 11 (1): 63–81.

Soper, Kate. 1999. The politics of nature: Reflection on hedonism, progress and ecology. *Capitalism, Nature, Socialism* 10 (2): 47–70.

Stolle, Dietlind, Marc Hooghe, and Michele Micheletti. 2005. Politics in the supermarket: Political consumerism as a form of political participation. *International Political Science Review* 26 (3): 245–69.

Talbot, John M. 2004. *Grounds for Agreement: The Political Economy of the Coffee Commodity Chain*. Boulder, CO: Rowman and Littlefield.

TransFair USA. 2010. Almanac 2009. Accessed September 6, 2010. http://transfairusa.org/pdfs/Almanac_2009.pdf.

UNDP (United Nations Development Programme). 2011. The human development concept. Accessed September 16, 2011. http://hdr.undp.org/en/humandev/.

Utting, Karla. 2009. Assessing the impact of Fair Trade coffee: Towards an integrative framework. *Journal of Business Ethics* 86: 127–49.

Utting-Chamorro. Karla. 2005. Does Fair Trade make a difference? The case of small coffee producers in Nicaragua. *Development in Practice* 15 (3/4): 584–99.

VanderHoff Boersma, Francisco. 2009. The urgency and necessity of a different type of market: The perspective of producers organized within the Fair Trade market. *Journal of Business Ethics* 86: 51–61.

VanderHoff Boersma, Franz. 2002. Poverty alleviation through participation in Fair Trade coffee networks: The case of UCIRI, Oaxaca, Mexico. Center for Fair and Alternative Trade. Accessed November 20, 2003. www.cfat.colostate.edu/research/.

Varul, Matthias Zick, and Dana Wilson-Kovacs. 2008. Fair Trade consumerism as an everyday ethical practice: A comparative perspective. Economic and Social Research Council, University of Exeter.

Walmart. 2010. Sustainability. Accessed July 28, 2010. http://instoresnow.walmart.com/Sustainability.aspx?povid=cat14503-env172199-module042610-lLink_wnsus.

Walsh, Jean Marie. 2004. Fair Trade in the fields: Outcomes for Peruvian coffee producers. Master's thesis, Massachusetts Institute of Technology.

Warde, Alan. 1994. Consumption, identity-formation and uncertainty. *Sociology* 28 (4): 877–98.

Weekes, Claire. 2008. Wine "one of the fastest growing" Fairtrade products. Harpers Wine and Spirit Trades Review, September 2. Accessed January 30, 2009. www.harpers.co.uk/news/news-headlines/7030-wine-one-of-fastest-growing-fairtrade-products.html.

Whole Foods Market. 2009a. Our whole trade guarantee. Accessed September 25, 2009. www.wholefoodsmarket.com/products/whole-trade.php.

———. 2009b. Whole Trade Certified partners. Accessed September 25, 2009. www.wholefoodsmarket.com/products/certifier-partners.php.

Wilkinson, John. 2007. Fair Trade: Dynamic and dilemmas of a market-oriented global social movement. *Journal of Consumer Policy* 30: 219–39.

Wilkinson, John, and Gilberto Mascarenhas. 2007. The making of the Fair Trade movement in the South: The Brazilian case. In *Fair Trade: The Challenges of Transforming Globalization*, edited by Laura T. Raynolds, Douglas Murphy, and John Wilkinson, 157–79. London: Routledge.

Wit, Arjaan, Henk Wilke, and Harmen Oppewal. 1992. Fairness in asymmetrical social dilemmas. In *Social Dilemmas: Theoretical Issues and Research Findings* edited by Wim B. Liebrand, David Messick, and Henk Wilke, 183–197. Oxford, UK: Perga.

Contributors

Rebecca Kahn works for an AIDS service organization, connecting clients with community resources that will allow them to preserve their housing. She has always been interested in and involved with food and trade justice, and studied Fair Trade coffee consumption for her master's thesis at the University of Auckland, New Zealand. She lives and works in Boston, Massachusetts.

April Linton is assistant professor of sociology at the University of California, San Diego. She studies transnational social movements, global development, and international migration. Her work on Fair Trade includes articles about NGO business alliances and corporate social responsibility in the coffee industry, and social transformation in the South African wine industry. She is coeditor of *The Global Governance of Food* (Routledge, 2009).

Marie Murphy is a PhD candidate in sociology at the University of California, San Diego. Her dissertation ethnographically examines the production of knowledge about sexuality within medical education. Marie spent three years as a Peace Corps volunteer in Zambia, where her experiences fueled her interest in Fair Trade and social justice in the global South.

Index

Page numbers followed by b, f, t, and n refer to boxes, figures, tables, and notes, respectively.